Automated Testing in Microsoft Dynamics 365 Business Central

Second Edition

Efficiently automate test cases for faster development cycles with less time needed for manual testing

Luc van Vugt

BIRMINGHAM—MUMBAI

Automated Testing in Microsoft Dynamics 365 Business Central
Second Edition

Copyright © 2021 Packt Publishing

Group Product Manager: Alok Dhuri
Publishing Product Manager: Richa Tripathi
Senior Editor: Ruvika Rao
Content Development Editors: Nithya Sadanandan and Ananya Endow
Technical Editor: Pradeep Sahu
Copy Editor: Safis Editing
Project Coordinator: Deeksha Thakkar
Proofreader: Safis Editing
Indexer: Subalakshmi Govindhan
Production Designer: Sinhayna Bais
Marketing Coordinator: Sonakshi Bubbar

First published: April 2019

Second edition: December 2021

Production reference: 1081221

Published by Packt Publishing Ltd.
Livery Place
35 Livery Street
Birmingham
B3 2PB, UK.

ISBN 978-1-80181-642-7

www.packt.com

To all that keep me growing as a human being; not in the least my wife, Saskia, our three great kids, Cyriel, Elvira, and Micha, my family and friends, and all the students I had the chance to teach and meet over the years.

– Luc van Vugt

Contributors

About the author

In 1999, **Luc van Vugt** stepped into the Dynamics 365 Business Central world, training hundreds of developers. After Microsoft acquired Navision, he joined Microsoft's Dynamics localization team GDL as a tester, UA specialist, and project lead. For 6 years, he was closely involved with all successive releases. Since he left Microsoft, he has become an active community member by means of his blog. He is a co-founder of the Dutch Dynamics Community and a speaker at conferences. For all his community work, Luc has been awarded MS MVP since 2011. In 2014, he co-founded NAV Skills, supporting Dynamics 365 Business Central, at that time still called Dynamics NAV, pros around the globe with, among other things, webinars until 2019. He has continued ever since to organize webinars under the flag of Areopa webinars. In 2012, Luc started fluxxus.nl, doing miscellaneous projects, including Dynamics 365 Business Central development courses and workshops on design patterns, and automated testing. Luc is one of the main driving forces on test automation in the Dynamics 365 Business Central community.

There are many people I am thankful to, for helping me gain the knowledge that I am sharing with you – but a couple to mention explicitly here:

The interview volunteers where the project of this second edition started with: Tin, Xavi, Maarten, Gert, Krzysztof, Jeremy, Guido, and Olaf.

Xavi Ametller Serrat, Maarten Gerritsen, and Daniel Rimmelzwaan for their invaluable reviews with great suggestions that gave the book more than I could have on my own.

The Packt team, who did a great job, surpassing my initial expectations by far.

About the reviewers

Daniel Rimmelzwaan was born and raised in the Netherlands and moved to the USA at the end of 1999 to be with his American wife in the beautiful mountains of Northern Arizona. Over the past two decades, Daniel has worked in just about every role around implementing NAV and now Business Central. He currently owns his own business (`risplus.com`) doing business analysis, architecture and design, and development. He is enjoying his career more than ever.

Ever since he started working with Business Central, Daniel has contributed to its online communities. For those contributions, Daniel received his first of seventeen consecutive Microsoft MVP awards in July 2005. Follow Daniel's blog at `http://thedenster.com/`.

Maarten Gerritsen has over 20 years of experience with Business Central and its predecessors, NAV/Navision. With Business Central, he can do what he loves most: creating solutions for the SMB market. He has a strong focus on process analysis and solution architecture. He worked for implementation partners and has been self-employed since 2007 (`www.intelligo.nl`).

Finally, when starting to develop for AppSource in 2019, he was *forced* to do something with automated testing. This got him hooked and he got passionate about the ATDD methodology, and it changed the way he creates solutions with Business Central.

Maarten lives in the Netherlands with his wife and two children. They teach him the most valuable lessons in life every day.

Xavier Ametller has been working in software development since 2003. Around 2009, he was a tester in what was called Microsoft Dynamics NAV, where he was lucky to meet Luc and discovered what now is Business Central.

When he left Microsoft, Xavi moved away from the BC world so he's not an expert on the topic. On the other hand, Xavi has participated in developing software in multiple roles and contexts during his career, so he tried to bring a view from the outside world while reviewing this book.

Xavi has a blog at `https://www.myowncommonsense.com/` where he shares rants on software development. As a side quest to his job, he offers training and consultancy on development-related topics.

Xavi and his family live happily in Catalonia, Spain.

Table of Contents

Section 2: Automated Testing in Microsoft Dynamics 365 Business Central

3

The Testability Framework

4

The Test Tools, Standard Tests, and Standard Test Libraries

Section 3: Designing and Building Automated Tests for Microsoft Dynamics 365 Business Central

5

Test Plan and Test Design

6

From Customer Wish to Test Automation – the Basics

7

From Customer Wish to Test Automation – Next Level

8

From Customer Wish to Test Automation – the TDD way

Section 4: Integrating Automated Tests in Your Daily Development Practice

9

How to Integrate Test Automation in Daily Development Practice

10

Getting Business Central Standard Tests Working on Your Code

Section 5: Advanced Topics

11

How to Construct Complex Scenarios

12
Writing Testable Code

13
Testing Incoming and Outgoing Calls

Section 6: Appendix

Appendix

Getting Up and Running with Business Central, VS Code, and the GitHub Project

Index

Other Books You May Enjoy

Preface

With recent changes, not least the evolution of Microsoft Dynamics NAV into Dynamics 365 Business Central, your development practices need to become more formal. Your success will depend, more than ever before, on your ability to determine the quality of your work more quickly and more consistently. Traditional, that is, manual testing, practices will not be sufficient anymore. So, you need to learn automated testing and how to efficiently include this in your daily work. In this book, you will learn how it will enhance your work, functionally and technically, and hopefully learn to love it.

Who this book is for

This book is for developers, development managers, testers, functional consultants, and power users working with Microsoft Dynamics 365 Business Central. Being a book on automated testing techniques, it caters to both the functional and technical aspects of Business Central development.

The reader is assumed to have a general understanding of Dynamics 365 Business Central as an application and how this application can be extended. Even though some sections will be dedicated to the coding of automated tests, it's not mandatory to be able to write code for Dynamics 365 Business Central. Overall, it sure gives the reader an advantage to be able to think as a tester.

What this book covers

Chapter 1, *Introduction to Automated Testing*, introduces you to automated testing: why you would want to use it, what it exactly entails, and when you should use it.

Chapter 2, *Test Automation and Test-Driven Development*, gives a short description of what Test-Driven Development (TDD) is and points out parts that can be of value to your daily development practices too.

Chapter 3, *The Testability Framework*, elaborates on how Microsoft Dynamics 365 Business Central enables you to run automated tests, and what this so-called Testability Framework actually is by describing its five pillars.

Chapter 4, The Test Tools, Standard Tests, and Standard Test Libraries, introduces the testing tool that resides in Dynamics 365 Business Central and allows you to run tests. Alongside that, we will discuss the standard tests and test libraries that Microsoft provides with Dynamics 365 Business Central.

Chapter 5, Test Plan and Test Design, discusses a couple of concepts and design patterns that allow you to conceive your tests more effectively and efficiently.

Chapter 6, From Customer Wish to Test Automation – the Basics, teaches you – and allows you to practice, based on a business case – how to get from a customer's requirement to the implementation of automated tests. In this chapter, you will make use of standard test libraries and techniques discussed in the previous chapters. The examples in this chapter will teach you the basics of headless and UI testing, and how to handle a positive-negative test.

Chapter 7, From Customer Wish to Test Automation – Next Level, continues the business case from *Chapter 6, From Customer Wish to Test Automation – the Basics*, and introduces some more advanced techniques: how to make use of shared fixtures, how to parameterize tests, and how to handle UI elements and hand over variables to these so-called UI handlers.

Chapter 8, From Customer Wish to Test Automation – the TDD Way, elaborates on how you can get your (test) coding done the Test-Driven Development way, putting test first all the way. Meanwhile you'll get introduced to two more examples and continue with the same business case from the previous two chapters: how to go about testing a report and how to set up permission testing.

Chapter 9, How to Integrate Test Automation in Daily Development Practice, discusses a number of best practices that might turn out to be beneficial for you and your team in getting test automation up and running in your day-to-day work.

Chapter 10, Getting Business Central Standard Tests Working on Your Code, discusses why you would want to use the standard test collateral Microsoft provides with Dynamics 365 Business Central, and how to fix errors when standard tests fail due to your extension of the standard application.

Chapter 11, How to Construct Complex Scenarios, shows you how to go about getting automated tests in place for complex scenarios, how to create reusable parts for them, and how to make use of already existing helper functions in the Microsoft test libraries.

Chapter 12, Writing Testable Code, introduces the concept of testable code: how to get your application code better suited to be tested with automated tests.

Chapter 13, Testing Incoming and Outgoing Calls, discusses how to test features that communicate with external systems, either being called from or calling on the outside world.

Appendix, Getting Up and Running with Business Central, VS Code, and the GitHub Project, pays attention to VS Code and AL development, and the code examples to be found in the repository on GitHub.

To get the most out of this book

This book is an introduction to test automation for Dynamics 365 Business Central. On one hand, various concepts and terminology are discussed, and on the other hand, we will also practice them by coding tests. To get the most out of this book, you might want to practice what is preached by implementing the code examples discussed. As this book, however, does not cover how to program for Business Central, you might first want to read the tips given in *Appendix, Getting Up and Running with Business Central, VS Code, and the GitHub Project*.

If your learning style is to start by finding out the principles, terminology, and concepts, start reading *Chapter 1*, and move slowly into the more practical *Section 3, Designing and Building Automated Tests for Microsoft Dynamics 365 Business Central*. If your style is more learning by doing, you could dare to take a deep dive into *Chapter 6*, *Chapter 7*, and *Chapter 8* straight away, and read the various backgrounds later or while working through the chapters.

> **Note**
> We advise you to type the code yourself or access the code from the book's GitHub repository (a link is available in the next section). Doing so will help you avoid any potential errors related to the copying and pasting of code.

Download the example code files

You can download the example code files for this book from GitHub at `https://github.com/PacktPublishing/Automated-Testing-in-Microsoft-Dynamics-365-Business-Central-Second-Edition`. If there's an update to the code, it will be updated in the GitHub repository.

We also have other code bundles from our rich catalog of books and videos available at `https://github.com/PacktPublishing/`. Check them out!

Get the color images

You can find the color images of the screenshots and diagrams used in this book in the GitHub repository: `https://github.com/PacktPublishing/Automated-Testing-in-Microsoft-Dynamics-365-Business-Central-Second-Edition/tree/main/Graphics`.

Conventions used

There are a number of text conventions used throughout this book.

`Code in text`: Indicates code words in text, database table names, folder names, filenames, file extensions, pathnames, dummy URLs, user input, and Twitter handles. Here is an example: "The only step to be taken is to add the following code to the relevant `Initialize` function."

A block of code is set as follows:

```
fields
{
  field(1; Code; Code[10]){}
  field(2; Description; Text[50]){}
}
```

When we wish to draw your attention to a particular part of a code block, the relevant lines or items are set in bold and in specific coloring with respect to certain tags:

```
[FEATURE] LookupValue Customer
[SCENARIO #0001] Assign lookup value to customer
[GIVEN] A lookup value
[GIVEN] A customer
```

Bold: Indicates a new term, an important word, or words that you see onscreen. For instance, words in menus or dialog boxes appear in **bold**. Here is an example: "Select **System info** from the **Administration** panel."

Tips or important notes
Appear like this.

Get in touch

Feedback from our readers is always welcome.

General feedback: If you have questions about any aspect of this book, email us at `customercare@packtpub.com` and mention the book title in the subject of your message.

Errata: Although we have taken every care to ensure the accuracy of our content, mistakes do happen. If you have found a mistake in this book, we would be grateful if you would report this to us. Please visit `www.packtpub.com/support/errata` and fill in the form.

Piracy: If you come across any illegal copies of our works in any form on the internet, we would be grateful if you would provide us with the location address or website name. Please contact us at copyright@packt.com with a link to the material.

If you are interested in becoming an author: If there is a topic that you have expertise in and you are interested in either writing or contributing to a book, please visit authors.packtpub.com.

Share Your Thoughts

Once you've read *Automated Testing in Microsoft Dynamics 365 Business Central*, we'd love to hear your thoughts! Scan the QR code below to go straight to the Amazon review page for this book and share
your feedback.

https://packt.link/r/1801816425

Your review is important to us and the tech community and will help us make sure we're delivering excellent quality content.

Section 1: Automated Testing – A General Overview

In this section, you will be introduced to automated testing. We will discuss why you would want to use it, what it exactly entails, and when you should use it.

This section contains the following chapters:

- *Chapter 1, Introduction to Automated Testing*
- *Chapter 2, Test Automation and Test-Driven Development*

1
Introduction to Automated Testing

A couple of years ago, I finally got to write this book on application test automation – one I had wanted to write for a long time already, as automated testing doesn't appear to be a *love-at-first-sight* topic for many. And, like testing in general, in many implementations, it tends to be subordinate to requirements specifications and application coding, be it a project or a product. And who really loves testing? It is not typically what the average developer gets enthusiastic about. Even functional consultants do not easily step up, for example, when raising this question during many of my workshops on automated testing.

So, when I started writing this book, inevitably, I did pose the question to myself: do I love testing? And I answered it with a yes. A big YES. This subsequently led to additional questions, such as: What makes me love testing? Did I always love testing? Why do I love it while the rest of the world seems not to? Questions with answers that made me go around the world evangelizing test automation, stubbornly sharing my findings, and pushing Microsoft to improve their tests to make them better and reusable. And all because I reckon that it is fun – BIG fun!

Having been an application tester in the former Dynamics NAV **Global Development Localization** (**GDL**) team at Microsoft, I surely got exposed to the testing virus. You could say that I had to learn to do the testing job, as I was paid for it. But it also suited me well, with me apparently having this specific DNA kind of a thing that makes testers what testers are. Having pride in breaking *the thing* and loving to prove its robustness (hopefully) at the same time. And, not in the least, daring to break it, running the risk that your developers will no longer like you.

One afternoon at Microsoft, my developer team member walked into our office and stopped next to my desk. Him being very tall and me sitting, I had to turn my head up to look at him.

"What's up?" I asked.

"Don't you like me anymore?" he responded.

Me: What?

Him: "Don't you like me anymore?"

Me: "Nope, still like you. Why?"

Him: "You rejected my code; don't you like me anymore?"

Me: "Dude, I still like you, but concerning your code, my tests show it somehow useless."

Testing is not rocket science, nor is automated testing. It's just another learnable skill. From a developer's perspective, however, it requires a change of mindset to write code with a totally different purpose than you are used to. And we all know that change is often not the easiest thing to achieve. Because of this, it's not unusual for attendees at my workshops to get to a certain level of frustration.

Application testing is a mindset, and it needs a big dose of discipline too – the discipline to do what needs to be done: to verify that the feature was built right; to verify that the feature under test meets the requirements – and the discipline when bugs are reported and fixed to execute the whole test run again and again with any new bug, to ensure that the verification is complete.

I tend to see myself as a disciplined professional. I have been quite a disciplined tester, with a high rate of bug reporting. But did I always love testing? You know, in those days, all our tests were executed manually, and with each bug I found, my discipline was challenged in some way. Imagine my mind running when executing the fifth test run after another bug fix. It's 4:00 P.M. and I am almost done, and the feature under test has to be delivered today. At breakfast, I promised my wife that I would be home on time, for whatever reason. Let's pick a random one: our wedding anniversary. So, me being a disciplined tester, my promise to be home on time, 4:00 P.M., and … I … hit … another … bug. Knowing that fixing it and rerunning the tests would take at least a couple of hours; how do you think my mind is running? Right: *binary*.

I had two options:

- Reporting the bug would keep me at work and make my wife highly disappointed, to say the least.

- Not reporting the bug would get me home on time and save me trouble at home.

Had automated tests been in place, the choice would have been quite simple: the first option, resulting in no hassle at home and no hassle at work. Welcome to my world. Welcome to test automation!

In this chapter, we will discuss the following topics:

- Why automated testing?

- When to use automated testing?

- What is automated testing?

> **Note**
>
> If you prefer to read the *what* first, you might first want to jump to *What is automated testing?*

Why automated testing?

Plainly said: it all boils down to saving you a lot of hassle. There are no emotions or time-intensive execution that will keep you from (re)running tests. It's just a matter of pushing the button and getting the tests carried out. This is easily reproducible. Each time you push the button, the test will be executed exactly the same. It is fast to execute, as automated tests are "light-years" faster than any of us could do, and objective: a test fails or succeeds, and this will be part of the overall reporting. No human emotions that could hide it away.

If it was as straightforward as that, then why haven't we been doing this in the Dynamics 365 Business Central world all these years? You probably can come up with a relevant number of arguments yourself. Of which the most prominent might be we do not have time for that. Or maybe: who's going to pay for that?

Why not?

Before elaborating on any arguments, pros or cons, let me create a more complete *why not?* list. Let's call it the *whys of non-automated testing*:

- Costs are too high and will make us uncompetitive.

- We are not used to doing it this way and our sales and management team don't see the benefits and cannot be "convinced."

- Dynamics 365 Business Central platform does not enable it.

- Customers do the testing, so why should we bother?

- Who's going to write the test code? We already have a hard time finding people.

- Our everyday business does not leave room to add a new discipline.
- There are too many different projects to allow for test automation.
- Microsoft has tests automated, and still, Dynamics 365 Business Central is not bug-free.

Why yes?

Dynamics 365 Business Central is not as simple as it used to be way back when it was called **Navigator**, **Navision Financials**, or **Microsoft Business Solutions – Navision**. And the world around us isn't as one-fold either. Our development practices are becoming more formal, and, with this, the call for testing automation is pressing on us for almost the same reasons as the *whys of non-automated testing*:

- Drive testing upstream and save costs.
- Dynamics 365 Business Central platform enables test automation.
- Relying on customers to do the testing isn't a great idea.
- Having a hard time finding people – start automating your tests.
- Test automation will free up time for everyday business.
- Keep on handling different projects because of test automation.

Let's check each of these aspects separately.

Drive testing upstream and save costs

Regarding costs, I tend to say that a Dynamics 365 Business Central project goes 25% over budget on average, mainly due to bug fixing after *go-live*. Quite a number of attendees of my workshops, however, have tended to correct me, saying that this is on the low side. Whatever percentage is the case, the math is quite simple. If you're spending 25% extra at the end of the line, why not push it upstream and spend it during the development phase on automated testing, and meanwhile build up reusable collateral? Even more, this also ends up being a powerful multiplier as any defect solved earlier in the project chain, or what we also call **upstream**, is a factor cheaper than being solved later in the project, that is, **downstream**.

If my memory is not failing me, this factor could be as large as 1,000. During my time at Microsoft in the 2000s, research was performed on the cost of catching a bug in the different stages of developing a major release of a product. Catching a bug after the release was found to be approximately 1,000 times costlier than when catching the bug already at requirement specification.

Translating this to the world of an **independent software vendor (ISV)**, this factor might roughly be 10 times lower. Meaning that the cost of catching a bug all the way downstream would be 100 times higher than all the way upstream. In the case of a **value-added reseller (VAR)** doing one-off projects, this factor probably is 100 times lower. So, whatever the factors would be, any spending upstream is more cost-effective than downstream, be it more formalized testing, better app coding, code inspection, or specifying more detailed requirements.

> **Note**
>
> A while ago, I wrote a joint blog post that also discussed this. You might want to read it:
>
> `https://www.fluxxus.nl/index.php/BC/from-a-testing-`
> `mindset-to-a-quality-assurance-based-mindset.`

Dynamics 365 Business Central platform enables test automation

In all honesty, this is a no-brainer as this is the topic of this book. But it is worthwhile realizing that the testability framework inside the platform has been there ever since version 2009 SP1, released in the summer of 2009. So, for over 10 years, the platform has enabled us to build automated tests. Does it sound strange if I say that we have been sleeping for a long time? Sticking to old habits? At least most of us.

Relying on customers to do the testing isn't a great idea

I agree that customers might know their features the best, and as such, they are the presumable testers. But can you rely 100% that testing isn't being squeezed between deadlines of implementation, and, in addition, between deadlines of their daily work? And still, in what way does their testing contribute to a more effective test effort in the future? How structured and reproducible will it be?

Posing these questions answers them already. It isn't a great idea, in general, to rely on customer testing if you want to improve your development practices. Having said that, this doesn't mean that customers should not be included; by all means, incorporate them into setting up your automated tests. We will elaborate on that more later.

Having a hard time finding people – start automating your tests

At this very moment, all implementation partners in the Dynamics world are having a hard time finding people to add to their teams in order to get the work done. Note that I deliberately didn't use the adjective *right* in that sentence. We all are facing this challenge. And, even if human resources were abundant, practices show that, with a growing business in either volume or size, the number of resources used does not grow proportionally.

Consequently, we must all invest in changing our daily practices, which very often leads to automation, using, for example, PowerShell to automate administrative tasks and RapidStart methodology for configuring new companies. Likewise, writing automated tests to make your test efforts easier and faster. Indeed, it needs a certain investment to get it up and running, but it will save you time in the end.

Test automation will free up time for everyday business

Similar to getting the job done with proportionally fewer resources, test automation will eventually be of help in freeing up time for everyday business. This comes with an initial price but will pay off in due time.

> **Note**
>
> A logical question posed when I touch on the topic of spending time when introducing test automation concerns the ratio of time spent on application and test coding. Typically, in the Microsoft Dynamics 365 Business Central team, this is a 1:1 ratio. Meaning that, for each hour of application coding, there is 1 hour of test coding.

Keep on handling different projects because of test automation

Traditional Dynamics 365 Business Central implementation partners typically have their hands full of customers with a one-off solution. And, due to this, they have dedicated teams or consultants taking care of these installations; testing is handled in close cooperation with the end user, with every test run putting a significant load on those involved. Imagine what it would mean to have automated test collateral in place – how you would be able to keep on servicing those one-off projects as your business expands.

On any major development platform, such as Visual Studio, it has been a common practice for a long time already that applications are delivered with test automation in place. Note that more and more customers are aware of these practices. And, more and more will ask you why you are not providing automated tests for their solution.

Each existing project is a threshold to take having a lot of functionality and no automated tests. In a lot of cases, a major part of the features used is standard Dynamics 365 Business Central functionality, and, for these, Microsoft has made their own tests available since version 11, that is, Dynamics NAV 2016. Altogether, over 33,000 tests have been made available for the latest version of Business Central. This is a humongous number, of which you might take advantage and make a relatively quick start on. We will discuss these tests and how you could get them running on any solution later.

Some more arguments

Still not convinced why you would/should/could start using test automation? Do you need more arguments to sell it inside your company and to your customers?

Here are some more arguments:

- Nobody loves testing.
- Reduced risks and greater satisfaction.
- Once the learning curve is over, it will often be quicker than manual testing.
- Leveraged development practice.
- Shorter update cycles and the AppSource paradigm.
- Test automation is required(?).

Nobody loves testing

Well, almost nobody. And, surely, when testing means rerunning and rerunning today, tomorrow, and next year, it tends to become a nuisance, which deteriorates the testing discipline. Automate tasks that bore people and free up time for more relevant testing where manual work makes a difference.

Reduced risks and greater satisfaction

Having automated test collateral enables you to have quicker insight than ever before into the state of the code. And, at the same time, when building up this collection, the regression of bugs and the insertion of new ones will be much lower than ever before. This all leads to reduced risks and greater customer satisfaction, and your management will love this.

Once the learning curve is over, it will often be quicker than manual testing

Learning this new skill of automating your tests will take a while, no doubt about that. But, once mastered, conceiving and creating them will often be much quicker than doing the thing manually as tests often are variations of each other. Copying and pasting with code is ... well ... can you do such a thing with manual testing? Not to mention to choose between re-running these automated tests or the manual tests.

Leveraged development practice

Test automation is not just automating your test practice and speeding up some parts of your daily work, it's opening up new possibilities, as obvious as they can be. Just to mention a few:

- Not having to manually (re)execute the obvious, ever-reappearing tests opens up room for manual testers to focus on challenging the system, to really try *to break the damn thing*.

- As automated tests can be executed by anybody or any automated process, a developer has a new tool at hand while developing the application code with fast feedback: with any change relevant, automated tests can be run to check whether no regression occurs.

- Test automation needs more details upfront, which will take some more time, but at the same time allows a much better, more meaningful discussion about the requirements; through the test definitions, the requirements are better *tested*.

> **Note**
> You might want to read my blog post for more examples of how your practice can be leveraged: `https://www.opgona.training/index.php/2021/02/25/test-automation-its-not-a-buzz-word-but-a-grand-helping-hand/`.

Shorter update cycles and the AppSource paradigm

With iterative methodologies – such as Scrum and Agile – and cloud services, it has become a normal practice that updates are delivered with shorter time intervals, leaving even less time to get full manual testing done, by customers and dedicated resources in your team. Clearly, Dynamics 365 Business Central is a part of this story and your app too. As it will also be a part of the AppSource paradigm, where companies around the world use Business Central in the cloud and are able to basically install any app from AppSource and start using it without any formal relation to and instruction from the manufacturer. If the app appears to be buggy, then they probably will just as easily uninstall it and go for another. You cannot afford for your app to be put aside because your testing practice does not fit this reality. In the **Software as a service (SaaS)** world, automated testing is a must to guarantee a quality product.

Test automation is required(?)

In the first edition, I started this section with:

Last but not least, on this incomplete discussion of arguments on the whys of test automation, and perhaps the sole reason for you reading this book: automated tests are required by Microsoft when you are going to sell your Dynamics 365 Business Central extension on AppSource.

This is no longer the case as Microsoft does not require automated testing as part of the AppSource validation anymore. They, however, *strongly recommend … using automated tests*, hence the question mark added to the section heading.

But even though Microsoft isn't forcing us anymore, know that our customers will be requesting this from us more and more, seeing that your competitors do practice and provide automated testing. The latter giving your competitor the advantage of having more time to add features and release more often.

> **Note**
>
> Read Krzysztof Bialowas' blog post where he discusses this Microsoft policy change with a reference to the Microsoft communication on this:
>
> `http://www.mynavblog.com/2021/02/16/automated-tests-dont-listen-to-microsoft/`.

Silver bullet?

Notwithstanding all the above, you might rightfully wonder whether test automation is a silver bullet that will solve everything. And I cannot deny that! However, I can tell you that, if exercised well, it will surely raise the quality of your development effort. As mentioned before, it has the following benefits:

- Easy to reproduce

- Fast to execute

- Objective

When to use automated testing?

Enough arguments to convince you *why* you would want to use automated tests, I guess. But how about *when* to use them? Ideally, this would be whenever code is changed to show that this functionality, which has already been tested, is still working as it should, to show that recent modifications have not compromised the existing application.

This sounds logical, but what does this mean when you have no automated tests in place? How do you go about starting to create your first ones? Basically, I would advise you to use the following criteria:

- What code change will give the highest return on investment when creating automated tests?

- For what code change will your test automation creation improve your test coding skills the most?

Using these two criteria, the following examples of code changes are typical candidates for your first efforts:

- After go-live bug fixing
- Buggy code
- Frequently modified code
- Business-critical code being changed
- Code with no or low dependencies
- Refactoring of existing code
- New feature development
- Microsoft updates

After go-live bug fixing

An after go-live bug reveals an omission in the initial test effort that can often be traced back to a flaw in the requirements. Frequently, it has a restricted scope, and, not the least important, a clear reproduction scenario. And by all means, such a bug should be prevented from ever showing its ugly face again.

Buggy code

You have this feature that keeps on troubling you and your customers. Bugs keep on popping up in this code and it never seems to stop. The most elementary thing you should start with is the *after go-live* bug fixing approach as previously discussed. But, even more importantly, use this code to get started on your first test suite.

Bugs are a particularly useful starting point because they usually provide the following:

- A defined expectation
- Steps to reproduce the scenario
- A clear definition of how the code fails

These are three important elements of automated testing, as you will learn in the next few chapters.

Frequently modified code

One of the basic rules of good code governance is that code should only be changed when it is going to be tested. So, if the code is modified frequently, then the consequence is that it will also be tested frequently. Automating these tests will give a good return on investment for sure.

Business-critical code being changed

Thorough testing should always be the case, but, given the circumstances, it is unfortunately not always doable. Testing changes made to business-critical code, however, should always be exhaustive; that is, try to cover all scenarios of those processes and define a test for each one of them in your automated testing. You can simply not afford any failures on them. Make it a point of honor to find even the two to five percent of bugs that statistics tell us are always there!

Refactoring existing code

Refactoring code can be nerve-racking: removing, rewriting, and reshuffling code. How do you know it is still doing the job it used to? Does it not break anything else? It certainly needs to be tested. But, when manually done, it is often executed after the whole refactoring is ready. That might already be too late, as too many pieces got broken. Before refactoring any code, grant yourself peace of mind and start by getting an automated test suite in place to prove its validity of the original code. To achieve this, define the business scenarios and create tests for each of those scenarios to prove that the current functionality works. With every refactor step you take, run the suite. And again, to show the code is still behaving the same. This way, refactoring becomes fun. We will elaborate on refactoring more later.

New feature development

Starting from scratch and creating a new feature, writing both test and application code, will be an undeniable experience. For some, this might be the ultimate way to go. For others, this is a *bridge too far*, in which case, all previous candidates are probably better ones. In *Section 3, Designing and Building Automated Tests for Microsoft Dynamics 365 Business Central*, we will take this approach and show you the value of writing test code alongside application code.

Microsoft updates

Incorporating any update from Microsoft, be it on-premises or in the cloud, your features must be (re)tested to prove they're still functioning as before. In case you do not have automated tests in place, begin creating them. Do this based on the analysis of the various changes and the risks they might entail in terms of introducing errors.

> **Note**
>
> Working on test automation for a new feature will give you the best return on investment. It will lead to a better quality of the code and thus prevent a lot of bugs after go-live/release. But it might be too big a threshold when you start with test automation. Choose one of the above-suggested candidates to start your test automation *battle*.

What is automated testing?

We discussed *why* you might want to automate your tests and *when* to do this, and more specifically, where to start. But we didn't give any thought to what automated testing is. So, let's do that before we conclude this chapter.

With automated testing, we address the automation of application tests, that is, scripting manual application tests that check the validity of features. In our case, these are the features that reside in Dynamics 365 Business Central. You might have noticed that we have been using somewhat different terms for it:

- Test automation
- Automated tests
- Automated testing

These all mean the same thing!!!

On one hand, automated testing is replacing manual, often exploratory testing. It's replacing those manual tests that are repeatable and automatable, and often no fun (anymore) to execute.

On the other hand, they are complementary. Manual testing will still contribute to raising the quality of a feature, making use of creative and experienced human minds able to find holes in the current test design. Automated testing might also include so-called **unit tests**. These are tests that verify the working of atomic units that altogether make up a feature. Typically, these units would be single, global AL functions – units that would never be tested manually. Ultimately, both manual and automated tests serve the same goal: to verify that the object under test meets the requirements and doesn't do anything outside of the requirements.

Notes

(1) What is exploratory testing? Check out the following link for more information: `https://en.wikipedia.org/wiki/Exploratory_testing`.

(2) More on unit and functional tests can be found at the following link: `https://www.softwaretestinghelp.com/the-difference-between-unit-integration-and-functional-testing/`. We will also elaborate more on unit tests in this book.

Some more notes on automated tests

Before we head off to the next chapter, I would like to add a couple of notes on automated tests that haven't been touched on so far.

Automated tests are code too

There is no sense in denying: automated tests are also code. It takes time to design, implement, and maintain them. Like application code, any line of test code has a probability of containing a bug. The challenge is to keep this to a bare minimum. You can achieve this by:

- Making the test design a part of the requirements and requirements review
- Reviewing your test code like you would want to review your application code
- Making sure that tests always end with an adequate number of verifications

And, please, like any source, put your tests under source code management. They're a part of your product. Make sure to reserve time for this in your planning.

Automated tests need maintenance

Over time, your application changes and therefore the tests that go with it, be it manual or automated tests. Any change in the application, if well covered by test scenarios, will reflect in a number of failing tests as these no longer fit. Even a simple change in the order of the application code can lead to one or more tests falling over.

Testing vs. checking

One of the reviewers of this book, Xavi Ametller Serrat, pointed out the difference between testing and checking. With testing, we investigate and learn, while checking is about confirming and verifying. The first is typically a human exercise, while the second is what we can automate. Given this insight, *automated testing* should rather be called *automated checking*. In this book, however, the first term will be used as this links more to a common notice.

> **Note**
> An interesting read on testing versus checking can be found here: `https://www.developsense.com/blog/2009/08/testing-vs-checking/`.

Summary

In this chapter, we discussed *why* you would want to automate your application tests, and *when* you might want to create and run them. And, we concluded with a short description of the *what* – what is automated testing?

In *Chapter 2*, *Test Automation and Test-Driven Development*, you will learn about the Test-Driven Development principles and how they can be applied to Dynamics 365 Business Central development that includes test automation.

2
Test Automation and Test-Driven Development

It's not uncommon when people talk about **test automation** (**TA**) that the term **Test-Driven Development** (**TDD**) is uttered in the same breath. And quite often they are used seemingly as synonyms:

$$TA = TDD$$

But does that make sense?

In this chapter, we will first investigate how one relates to the other and whether they are in fact equal, followed by a more detailed description of what TDD is. After a short discussion on the benefits, we will illustrate how TDD could be applied in Microsoft Dynamics Business Central development. After that, we will discuss to what extent TDD comes into play when you go down the road of test automation.

This chapter covers the following main topics:

- TA versus TDD
- What is TDD?
- TDD – taking small steps
- TDD – the benefits
- TDD and Microsoft Dynamics 365 Business Central
- TDD – inside-out or outside-in

Where in the previous chapter we talked about the *what*, *why*, and *when* of test automation, this chapter will be a first step into the *how* of test automation.

TA versus TDD

When discussing TDD in more detail, it will become clear that **TDD** is highly dependent on **test automation** and will lead to comprehensive test collateral that covers the verification of your application code. As such it's a means to get both your test automation and application code in place. Or turning it around: test automation is part of the goal of TDD. But, as you will see below, TDD is more than that: it's a programming technique or, if that describes it better for you, another tool to shape your development practice. TDD and test automation do relate, but:

$$TA \neq TDD$$

Then you might wonder if TDD is the only way to get test automation in place and my short answer would be: "*No*," immediately followed by "*but it might make a lot of sense to use it.*" Read on!

What is TDD?

The shortest possible description of **TDD** is actually the term itself. It describes that, by using this as your methodology for your application development, this development will be driven by tests. Meaning, tests need to be defined to trigger the writing of your application code, and in this, your tests are directly derived from the requirements. The ultimate consequence of this is:

"No tests, no code."

You will never build code that does not have tests to it, and, as the tests are one-to-one related to the requirements, your application code will not have any undocumented features.

Only two rules to the game

The preceding description is not the basic definition of TDD. TDD is defined by only two rules, and everything else is deduced from that:

- **Rule 1**: Never write a single line of code unless you have a failing automated test.
- **Rule 2**: Eliminate duplication.

The first rule clearly defines the trigger for writing application code. If there is no failing automated test there is no reason to write app code. First you write the test, then the app code. Or shortly phrased: *no tests, no code*.

The second rule instructs you to eliminate any code duplication that emerged when fulfilling the first rule with the goal to write *reusable*, *readable*, and *minimalistic* code.

Having these two rules as a starting point how does this work out in practice? Let's check that next.

TDD – the red-green-refactor mantra

The *actionable steps* derived from the two rules are what has become known as the **red-green-refactor** mantra. Red and green respectively refer to a failing (red) and, eventually, succeeding (green) test.

The basis to start from is the so-called **test list**: a list of test scenarios derived from the requirements defining how to verify the behavior of the application code that will implement these requirements. Given the **test list**, the *red-green-refactor* mantra tells you to take the following steps, implementing one test at a time:

1. Take a test from the test list and write the test code.
2. Compile the test code yielding red as the application code is not yet there.
3. Implement *just enough* application code to make the test code compile.
4. Run the test seeing it probably fail, still red.
5. Adjust the application code *just enough* to make it pass, that is, green.
6. If applicable, **refactor** your code (either test or application code or both), one after the other to remove duplication and bring clarity in your code and rerun the test after each change to eventually prove all code is still well (green).
7. Move to the next test on the list and repeat from *step 1*.

The accompanying schema illustrates the *red-green-refactor* mantra in a flow chart way:

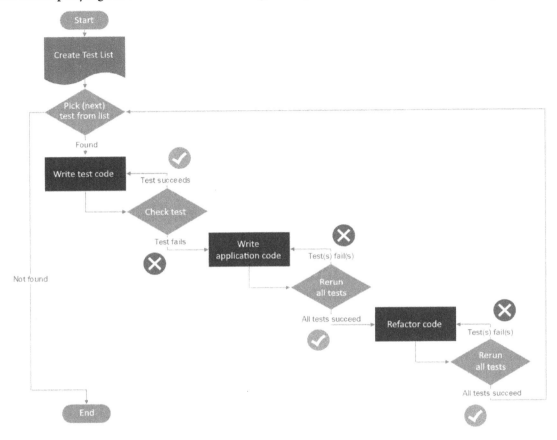

Figure 2.1 – The red-green-refactor mantra in flowchart format

The *red-green-refactor* mantra urges you to get things done step by step as efficiently, that is, as minimalistically, as possible. Only implement what is needed to get the test passed with one thing at a time, and as quickly as possible. Taking this mantra seriously, your first test might end up being of a somewhat ridiculous level, finding yourself hardcoding results (see the next section on *Fake it*). However, given that the test list has sufficient details, that is, consists of an adequate number of tests, implementing the next test will bring more detail to your code. This is called **triangulation**, which we will discuss in the next section.

Note

Often, when implementing requirements, we find out that they are not as complete as thought when reviewed and agreed upon. Thus, we find ourselves adding new details to them. Within the **TDD** context, this means that the **test list** is extended. Adding tests to this list is a valid part of TDD allowing you to adjust to new insights.

TDD – taking small steps

The success of applying TDD when coding your application depends highly on how small you dare make the steps you take. Each step should be *effective and efficient*. Don't do more than required for that step. Don't be tempted to start generalizing your code if it's not needed to get it working. Be aware that the second rule will be your guide here. Only in the case of code duplication, code generalization will happen, as rule two tells you to eliminate it.

Efficient and effective coding boils down to shifting between three modes of implementation that Kent Beck points out in his ground laying book, *Test-Driven Development – By Example* (see the section *Further reading*):

- **Fake it**
- **Use obvious implementation**
- **Triangulation**

Fake it

With the *fake it* implementation, also known as *fake it 'till you make it*, you focus on getting a test pass quickly, whatever it takes. Typically, you do know what your test is verifying and what the response of your application should be, but you do not yet know how this obviously should be coded. So, if a hardcoded value does the trick, that is, makes the test pass, that's what you'll use. As Kent Beck puts it with respect to the implementation of a function, make it…

"…return a constant and gradually replace constants with variables until you have the real code."

Or like I wrote above,

"…only implement what is needed to get the test pass … implementing the next test will bring more detail to your code."

Use obvious implementation

In case the implementation of the application code is obvious, there is clearly no need to beat around the bush. No reason to *fake it*, just implement it. But bear in mind, and dare to act on it, that even though it might look obvious, do not spend too much time on it, if the *fake it* implementation will get you much faster to `green`. The next mode, triangulation, will head you to the real code.

Triangulation

The use of the term **triangulation** refers to the surveying method of determining an unmeasured location, either on the earth's surface or in space, given two known locations and forming a triangle. Given these two known locations and the shape of the triangle, that is, its angles at the given locations, we are able to determine the third location. Using this method, old Greek scientists and probably their forerunners in Egypt, Mesopotamia, China, India, or in whatever other old culture, were able to make a calculation of the distance to the moon and sun.

Applying triangulation within the TDD context leads you from a *fake it* implementation to a meaningful algorithm: create two knowns, in parallel to the survey locations, by implementing, for example, a test for an error case and one for the simplest non-error case using a *fake it* implementation. Next, explore the unknown with a third, non-error, test case. So, starting from specific examples and adding more examples, following the TDD principles, you will end up with generic code that works for all of those examples. This way your code goes from specific to generic.

A simple, but very illustrative, example is the TDD implementation of a function that calculates the surface area of a square. Have a look at this nice post on DevOps Zone by Grzegorz Ziemoński: `https://dzone.com/articles/three-modes-of-tdd`.

TDD – the benefits

The benefits of putting TDD into practice are:

- Fewer bugs as you end up with code that has been tested already

- Only implement what is required

- Accompanying automated test collateral that allows you to run and check the sanity of the application code over and over again, now and in the future

- Little to no need for debugging; when code fails, return to the previous state of code (using source code management) and restart with smaller steps

- More pleasure in refactoring your application code with less stress as the tests safeguard its validity

- More meaningful reflection on the requirements as failing tests often point to flaws in the requirements

- For sure, an overall better product

Supposedly, you will have – next to the application code – more code due to the test code . With pre-TDD practices in mind, you might, however, surprisingly end up with less code as your application code only implements what has been asked for, at the same time has been implemented *efficiently and effectively*, and needed fewer corrections to the code due to bugs that were left in the final code.

> **Note**
>
> As noted, you will for sure have an overall better product, and potentially against a lower cost as Matej Kminiak's thesis *Using the method based on software development driven by tests in ERP implementations* shows, based on his work at Navertica, Czechia: `https://is.muni.cz/th/l5ica/`.

TDD and Microsoft Dynamics 365 Business Central

Now that we have talked about what TDD is, how it works, and what its benefits are, it makes sense to have a look at whether we could approach Microsoft Dynamics 365 Business Central development with TDD, or in a TDD-ish way. We will start this section with a general question and continue with a small example of how this would work out.

Is TDD in Business Central possible?

In the first edition of this book, I deliberately did not pay attention to TDD in the main part of the book, only to make some notes in the *Appendix*. I chose to do so as I knew TDD did not always resonate well and, as I wrote then…

…I didn't want anyone getting blocked by their knowledge of TDD, with it being a proven methodology, but also surrounded with skepticism.

I continued with a condensed description of TDD and concluded, relating it to all the examples worked out in the book (see further on in this second edition), that if I would have been applying TDD principles, the examples wouldn't have been very different, as with each example the:

- Requirements were converted into a test (on the test list)
- Test was coded step by step, using either of the three methods: fake it, obvious implementation, or triangulation
- Test was run yielding **red**, leading to some code adjustment, or yielding **green**, moving to the next test
- Code could have been refactored to remove duplication; within the constraints of the first edition this was, however, not part of the book, but only provided in the final code

Basically, this indicates that TDD can be used for Business Central development, but there are a couple of implicit hurdles to be taken. My experience says that these hurdles relate to the following:

- Requirements that are not always well suited to allow you to convert them efficiently into a test list.

- TDD is counterintuitive for many because they used to go and implement the application code straight away and testing would follow later; it's hard to change habits.

- *Upstream* time spent on detailing tests and writing both test and application code appears to be more as they have to be budgeted whereas *downstream* (that is, after go-live) time spent on bug fixing very often isn't.

- The nature of Business Central implementations, which most often do not start from scratch and have a broad context, both application and data related.

All in all, the full incorporation of TDD in our daily practices is a paradigm change. This isn't happening by itself and definitely needs time and mental investment. If done with commitment, it will be worthwhile as it both makes test automation a part of your implementation cycle and will improve the quality of requirements, code, and the product as a whole. As we will see further on in this book, working on test automation will implicitly make use of various TDD principles (where applicable, this will be denoted).

Don't know how to start?

Probably the simplest way to start practicing/exercising TDD is solving bugs following the next steps. Don't do anything else, just treat bugs like this, and you will start to understand how it works. Then start thinking about how to use it for new features:

1. Get a detailed enough reproduction description of the bug.
2. Based on this, you build the test (code).
3. Running the test should lead to red, as the bug is still there.
4. Fix the application code to make the test pass: green.
5. Refactor the bug fix and/or the automated test that now goes with it.

Very likely this is not far off from what you have been doing so far when reproducing bugs.

TDD in Business Central – by example

Now let's get started and show how TDD works out in Business Central. But before we do a technical note.

Where the example below illustrates that TDD can be applied in Business Central development, its primary focus is not the coding itself. As such I do not discuss the full code context, including the VS Code setup. However, the code can be found on the GitHub repo that comes with this book and is discussed in *Appendix, Getting Up and Running with Business Central, VS Code, and the GitHub Project*.

> **Note**
>
> In this chapter, we will not always make all parts of the code explicit, like variable declarations, test codeunits, and function constructions. We rather focus on the TDD process in the context of Business Central. As of *Chapter 3, The Testability Framework*, we will, however, include the full code context.

Example – dynamically hiding the totals sub frame on a sales document

The example we take has a simple requirement:

When the VAT business posting group of a sales document (header) boils down to a 0% VAT calculation the totals sub frame should be hidden, irrespective of the value of the VAT product posting group on any of the lines.

Before we start to work on the test list, I would like to share a couple of clarifying notes:

- In the context of a Cronus w1 demo company, this applies to the VAT business posting group **EXPORT**, where the VAT posting setup related to it always leads to a 0% VAT calculation as can be seen in the following screenshot.

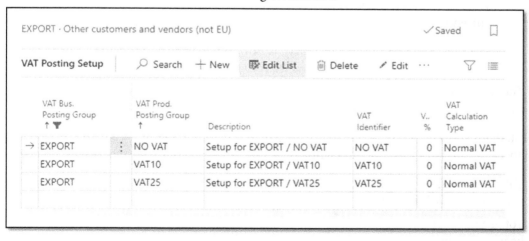

Figure 2.2 – VAT posting setup for VAT business posting group EXPORT

- The next screenshot shows a sales invoice. The totals sub frame is marked by the red rectangle. The sales invoice in this screenshot does not have the posting group **EXPORT**, since the VAT total is not zero.

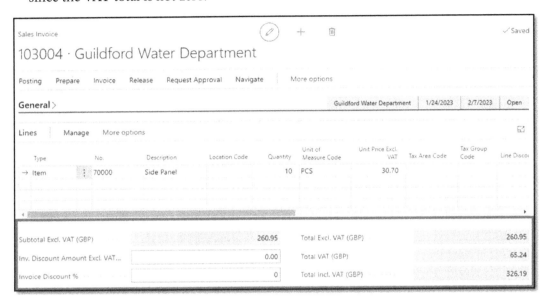

Figure 2.3 – The sales invoice page with totals sub frame marked

- If you're wondering about the validity of the requirement, realize it's just a made-up requirement to allow us to demonstrate how to implement a requirement through TDD.

From the requirement, we can derive our test list holding the following tests:

1. The totals sub frame on sales invoice page is **not** shown when VAT business posting group on the header equals a value that *always leads to 0 % VAT calculation*.

2. The totals sub frame on sales invoice page is shown when VAT business posting group on the header equals a value that *does not always lead to 0 % VAT calculation*.

As pointed out above, these tests could initially make use of the **EXPORT** VAT business posting group that exists in Cronus w1 demo company. A **fake it** version of these tests would then be:

1. Totals sub frame on sales invoice page is **not** shown when VAT business posting group on the header *equals* **EXPORT**.

2. Totals frame on sales invoice page is shown when VAT business posting group on the header *does not equal* **EXPORT**.

> **Note**
>
> You might have noticed that I emphasized the word *not*. I started doing this in my documentation some time ago as often a negation is overlooked, which could have dramatic consequences.

Test 1 (fake it): Totals sub frame on sales invoice page is not shown when VAT business posting group on header equals EXPORT

Let's start and practice the **red-green-refactor mantra** and first code the test.

- **Step 1 – Take a test from the test list and write the test code**: For this, we will need to implement the following:

  Have a sales invoice created with the VAT business posting group set to EXPORT:

  ```
  SalesHeader."Document Type" :=
    "Sales Document Type"::Invoice;
  SalesHeader.Insert(true);
  SalesHeader.Validate("VAT Bus. Posting Group", 'EXPORT');
  SalesHeader.Modify();
  ```

  Open the sales invoice document page for that sales invoice:

  ```
  SalesInvoice.OpenView();
  SalesInvoice.GoToRecord(SalesHeader);
  ```

  Verify that the sub frame, by means of the field Total Amount Excl. VAT, is *not* shown, that is, if it's shown throw an error:

  ```
  if SalesInvoice.SalesLines.
    "Total Amount Excl.VAT".Visible()
  then
    Error('%1 should NOT be visible.',
      SalesInvoice.SalesLines."Total Amount Excl.
        VAT".Caption);
  ```

  Close the page:

  ```
  SalesInvoice.Close();
  ```

- **Step 2 – Compile the test code yielding red as the application code is not yet there**: In this specific case, this does not apply as all code referenced (including tables, fields, and pages) do already exist, thus the test code compiles.

- **Step 3 – Implement just enough application code to make the test code compile**: No need to at this point. See the previous step.

- **Step 4 – Run the test seeing it probably fail, still red**: Running the test will produce red as we have not implemented any application code yet and the frame will be shown.

- **Step 5 – Adjust the application code just enough to make it pass, that is, green**: Time to add the application code. We want to take control of the visibility of the totals frame on the *Sales Invoice Subform* page. Therefore, we need to implement a page extension to change the `Visible` property of the page control `Control39` as follows.

> **Note**
>
> The sub frame does not have a specific name but is just called `Control39`. The platform, however, does not enable us to access the sub frame itself, instead, we will access one of the fields on the frame called `Total Amount Excl. VAT` as that will be hidden or shown with the sub frame.

Define a global variable to steer the `Visible` property of the frame:

```
layout
{
  modify(Control39)
  {
    Visible = SubFrameVisible;
  }
}

var
   SubFrameVisible: Boolean;
```

Create a function that determines whether the sub frame should be shown or not; note that the *fake it* part of the implementation of this test is in the hardcoded value EXPORT, against which we check the VAT Business Posting Group field of the sales header:

```
local procedure SetShowLinesSubFrame() Result: Boolean
var
   SalesHeader: Record "Sales Header";
   DocumentTypeFilter, NoFilter : Text;
```

```
begin
  Rec.FilterGroup(2);
  DocumentTypeFilter := Rec.GetFilter("Document Type");
  if (DocumentTypeFilter <> '') and
    (Rec.GetRangeMin("Document Type") =
      Rec.GetRangeMax("Document Type"))
  then begin
    Rec.FilterGroup(4);
    DocumentNoFilter := Rec.GetFilter("Document No.");
    if (DocumentNoFilter <> '') and
      (Rec.GetRangeMin("Document No.") =
        Rec.GetRangeMax("Document No."))
    then begin
      SalesHeader.Get(Rec."Document Type",
        Rec."Document No.");
      Result :=
        SalesHeader."VAT Bus. Posting Group" <> 'EXPORT';
    end;
  end;
  Rec.FilterGroup(0);
end;
```

Set SubFrameVisible:

```
trigger OnAfterGetCurrRecord()
begin
  SubFrameVisible := SetShowLinesSubFrame()
end;
```

Running the test will now produce **green** as the sub frame will *not* be shown.

- **Step 6 – Refactor your code**: Either test or application code or both, one after the other, and rerun the test after each change to prove all code is still well (green). When only the first test has been implemented, refactoring is not always obvious to do. So, let's move on.

- **Step 7 – Move to the next test on the list and repeat from Step 1**: Indeed, that is what we will do.

Test 2 (fake it): Totals frame on sales invoice page is shown when VAT business posting group on the header does not equal EXPORT

With the first test implemented, this one will be much easier as it contains a lot of similarities:

- **Step 1 – Take a test from the test list and write the test code**: For this, we will need to implement the following:

  Have a sales invoice created with the VAT business posting group not set to EXPORT; from Cronus, we will pick the DOMESTIC VAT business posting group having a setup with VAT % unequal to 0:

  ```
  SalesHeader."Document Type" :=
      "Sales Document Type"::Invoice;
  SalesHeader.Insert(true);
  SalesHeader.Validate(
      "VAT Bus. Posting Group", DOMESTIC);
  SalesHeader.Modify();
  ```

  Open the sales invoice document page for that sales invoice:

  ```
  SalesInvoice.OpenView();
  SalesInvoice.GoToRecord(SalesHeader);
  Verify that the sub frame, or actually the field Total Amount
  Excl. VAT, is shown, that is, if it's not shown throw an
  error:
  if not SalesInvoice.SalesLines.
      "Total Amount Excl. VAT".Visible()
  then
      Error('%1 should be visible.',
          SalesInvoice.SalesLines."Total Amount Excl.
          VAT".Caption);
  ```

  Close the page:

  ```
  SalesInvoice.Close();
  ```

  Compile the test code yielding red as the application code is not yet there.

- **Step 2 – Compile the test code yielding red as the application code is not yet there**: This step is not applicable as the application code we implemented above also applies here. We're just implementing a variation of the same test, so, we can jump to *step 5*.

- **Step 5 – Adjust the application code just enough to make it pass, that is, green**: Running the test will now produce green as the sub frame will be shown.

- **Step 6 – Refactor your code**: Either test or application code or both, one after the other, and rerun the test after each change to prove all code is still well (**green**). We haven't changed anything in the application code, but we did add a second test and with it the smell of duplication as it resembles the first test. Observe that both tests create a new sales header of type invoice.

The only difference is the value assigned to the VAT Business Posting Group field. Let's add the following local helper function to replace the duplicate code:

```
local procedure CreateSalesInvoice(
  var SalesHeader: Record "Sales Header";
  VATBusPostingGroup: Code[20])
begin
  SalesHeader."Document Type" :=
    "Sales Document Type"::Invoice;
  SalesHeader.Insert(true);
  SalesHeader.Validate(
    "VAT Bus. Posting Group", VATBusPostingGroup);
  SalesHeader.Modify();
end;
```

Open the *Sales Invoice* page, select the newly created sales invoice, and verify the visibility of the frame (or as a matter of fact, a field on the frame). For this, we can create another new helper function:

```
local procedure
  OpenSalesInvoicePageAndVerifyVisibility(
    SalesHeader: Record "Sales Header";
    FrameVisible: Boolean)
var
  SalesInvoice: TestPage "Sales Invoice";
begin
  SalesInvoice.OpenView();
  SalesInvoice.GoToRecord(SalesHeader);
  case FrameVisible of
    true:
      if not SalesInvoice.SalesLines.
          "Total Amount Excl. VAT".Visible()
      then
        Error('%1 should be visible.',
```

```
        SalesInvoice.SalesLines.
          "Total Amount Excl. VAT".Caption);
      false:
        if SalesInvoice.SalesLines.
          "Total Amount Excl. VAT".Visible()
        then
          Error('%1 should NOT be visible.',
            SalesInvoice.SalesLines.
              "Total Amount Excl. VAT".Caption);
    end;
    SalesInvoice.Close();
  end;
```

Now, call both new helper functions in both test functions to replace the duplicate code:

```
[Test]
procedure Test1()
var
  SalesHeader: Record "Sales Header";
  SalesInvoice: TestPage "Sales Invoice";
begin
  CreateSalesInvoice(SalesHeader, 'EXPORT');
  OpenSalesInvoicePageAndVerifyVisibility(
    SalesHeader, false);
end;

[Test]
procedure Test2()
var
  SalesHeader: Record "Sales Header";
  SalesInvoice: TestPage "Sales Invoice";
begin
  CreateSalesInvoice(SalesHeader, 'DOMESTIC');
  OpenSalesInvoicePageAndVerifyVisibility(
    SalesHeader, true);
end;
```

Rerunning the tests will yield **green** for both of them.

Note that it appears that we did include two refactoring changes in one step and only after these two steps ran the tests again to see them succeed. While developing this example, however, I did rerun the tests directly after introducing the first helper function and only added the second helper function to replace the remaining duplicate code after that (and ran the tests once more).

- **Step 7 – Move to the next test on the list and repeat from Step 1**: Well, we'd like to, but not yet, as our application code still is a *fake it* version and does need refactoring to get the real implementation.

We could have done it as part of the previous step but will execute this in the next section.

Refactor the application code to remove the fake it implementation

As said, the *fake it* implementation is embedded in the application code. It's in the highlighted statement in the following code:

```
local procedure SetShowLinesSubFrame() Result: Boolean
var
  SalesHeader: Record "Sales Header";
  DocumentTypeFilter, NoFilter : Text;
begin
  Rec.FilterGroup(2);
  DocumentTypeFilter := Rec.GetFilter("Document Type");
  if (DocumentTypeFilter <> '') and
     (Rec.GetRangeMin("Document Type") =
      Rec.GetRangeMax("Document Type"))
  then begin
    Rec.FilterGroup(4);
    DocumentNoFilter := Rec.GetFilter("Document No.");
    if (DocumentNoFilter <> '') and
       (Rec.GetRangeMin("Document No.") =
        Rec.GetRangeMax("Document No."))
    then begin
      SalesHeader.Get(Rec."Document Type", Rec."Document No.");
      Result :=
        SalesHeader."VAT Bus. Posting Group" <> 'EXPORT';
    end;
  end;
  Rec.FilterGroup(0);
end;
```

If we leave it as is, our code will fail in the following cases:

- Other VAT business posting groups, for which a 0% VAT calculation always applies and are selected on the sales invoice, will not be tested and will lead to the sub frame shown on the sales invoice page.

- Changes made to the setup of the VAT business posting group EXPORT, so a 0% VAT calculation no longer applies, will produce a wrong result.

So, instead of the comparison SalesHeader."VAT Bus. Posting Group" <> 'EXPORT', we need to determine whether or not a 0% VAT calculation always applies for the VAT business posting group selected. This can be achieved with the following function:

```
local procedure
  DoesZeroPercentVATCalculationAlwaysApply(
    VATBusPostingGroup: Code[20]): Boolean
var
  VATPostingSetup: Record "VAT Posting Setup";
begin
  VATPostingSetup.SetRange("VAT Bus. Posting Group",
    VATBusPostingGroup);
  VATPostingSetup.SetFilter("VAT %", '<>%1', 0);
  exit(VATPostingSetup.IsEmpty());
end;
```

The above highlighted statement line in the SetShowLinesSubFrame method will then become:

```
Result :=
  not DoesZeroPercentVATCalculationAlwaysApply(
    SalesHeader."VAT Bus. Posting Group");
```

Will the tests be successful? Well, let's run them again and see: **green. Green**!

Now that both tests have been rightfully implemented, we have completed our test list.

But as one of the reviewers, Maarten Gerritsen, validly noted: if there isn't any VAT posting set up for the VAT business posting group setup, then the sub frame will be hidden. Is this desired? As this isn't addressed by the requirement, this is an extension of it and as such a – nice example of a – new test to be added to the test list:

1. The totals sub frame on the sales invoice page is **not** shown when the VAT business posting group on the header equals a value that *always leads to a 0% VAT calculation*.

2. The totals sub frame on the sales invoice page is shown when the VAT business posting group on the header equals a value that *does not always lead to a 0% VAT calculation.*

3. **The totals sub frame on the sales invoice page is shown or not shown – to be decided still – when there is no VAT posting set up for the VAT business posting group on the header.**

With the implementation of the first two tests on the test list of this example, we demonstrated how the TDD approach can be used in implementing a requirement in Business Central – and therefore there's no need to continue with the third test. One of the major differences between this example (and potentially almost any requirement implementation with TDD in Business Central) and standard TDD examples is the data that needs to be provided. As mentioned before, this is a part of the nature of a Business Central implementation. We will experience the same in most of the examples worked on in this book.

> **Note**
>
> The full and completed code of this example can be found on the GitHub repo that accompanies this book:
>
> ```
> https://github.com/PacktPublishing/Automated-Testing-in-
> Microsoft-Dynamics-365-Business-Central-Second-Edition/
> tree/main/Chapter 02.
> ```
>
> Details on how to use this repository and how to set up VS Code are discussed in *Appendix, Getting Up and Running with Business Central, VS Code, and the GitHub Project.*

TDD – all the way?

I think I showed that, yes, with Business Central development you could do TTD all the way. But do you have to? No. As I stated in the first section of this chapter, *TA versus TDD*, TDD is not the only way to get test automation, "but it might make a lot of sense to use it," or at least parts of it. As you will see in *Chapter 6, From Customer Wish to Test Automation – the Basics*, and *Chapter 7, From Customer Wish to Test Automation – Next Level*, we will not make direct use of TDD. But as we will discuss in *Chapter 8, From Customer Wish to Test Automation – the TDD way* – where we will go the TDD way – there are a lot of TDD principles implicitly exercised in *Chapters 6* and *7*.

TDD – inside-out or outside-in

There is a lot more to be said and learned about TDD. In the course of writing this second edition, I had the opportunity of interviewing a number of peers on the book, and test automation in general. One of my interviewees – who also became one of the reviewers – was one of my former Microsoft test teammates, Xavi Ametller Serrat. Xavi continued his testing – and programming – career outside Microsoft, in a domain other than Business Central. During the interview, he pointed to many online resources on TDD, discussing things I had only touched upon so far: TDD **inside-out** or **outside-in**, two concepts that give insights on how TDD can be approached differently.

In her article *TDD – From the Inside Out or the Outside In?* (`https://8thlight.com/blog/georgina-mcfadyen/2016/06/27/inside-out-tdd-vs-outside-in.html`) Georgina McFadyen nicely describes these two approaches – by example – with their cons and pros. Jim Newbery signals us in short how we can mingle both approaches, and might suffer, in his post *Are you Detroit- or London-school?* (`https://tinnedfruit.com/list/20181004`), and in the meantime teaches us what Detroit- and London-school are about.

- **Detroit-school** (or **inside-out**) is bottom-up, developing units, one piece at a time, and does not need a prior full understanding of the system design.
- **London-school** (or **outside-in**) is top-down, developing the "whole," breaking down into smaller components when the need is there and mocking their real implementation to move on quickly, and thus deferring their real implementation till later.

Knowing Business Central implementations with their more functional, and thus more "holistic," approach you might have recognized yourself more in the London school, of which **Acceptance Test-Driven Development** (**ATDD**) is – indirectly – derived. We will introduce ATDD in *Chapter 5, Test Plan and Test Design*, and use it throughout the rest of the book as our test case design pattern.

For more, see the *Further reading* section below.

Summary

We started this chapter by discussing whether **TA** and **TDD** are synonyms. Where TDD makes use of test automation, it's evidently not a synonym of test automation but a very powerful means to get test automation in place. TDD leads you step by step from a test list, reflecting the requirements, through coded tests to application code. The so-called **red-green-refactor** mantra describes these steps. We pointed out that TDD is also about taking small steps to get tests passing efficiently and effectively, by making use of the three modes of TDD: fake it, obvious implementation, and triangulation. After this introduction to TDD, we shed our light on using TDD in Business Central concluding, based on arguments and an example, that TDD indeed can be used in this context even though a number of hurdles might have to be crossed.

With this chapter, we conclude the first section of this book, in which we introduced you to test automation in general and the specific methodology called TDD. In the next section, we will step into the general concepts of test automation in Business Central.

Further reading

Test-Driven Development has been around for quite a while and there has been a lot written on it in books and blog posts, so let's share some of them, starting with two masterpieces a lot of TDD-based programming and discussions have their roots in:

- Kent Beck, *Test-Driven Development – By Example*, Addison-Wesley, 2003

- James W. Newkirk and Alexei A. Vorontsov, *Test-Driven Development in Microsoft .NET*, Microsoft Press, 2004

- Triangulation:

  `https://www.fluxxus.nl/index.php/bc/tdd-in-nav-triangulation/`

  `http://codemanship.co.uk/parlezuml/blog/?postid=1157`

- A TDD example on Dynamics NAV:

  `https://www.fluxxus.nl/index.php/BC/test-driven-development-in-nav-intro`

- TDD Inside-out or Outside-in:

  `https://www.codurance.com/publications/2015/05/12/does-tdd-lead-to-good-design`

  `https://github.com/testdouble/contributing-tests/wiki/Detroit-school-TDD`

  `https://github.com/testdouble/contributing-tests/wiki/London-school-TDD`

  `https://www.browserstack.com/guide/tdd-vs-bdd-vs-atdd`

Section 2: Automated Testing in Microsoft Dynamics 365 Business Central

In this section, we will discuss how Microsoft Dynamics 365 Business Central enables you to run automated tests by means of the Testability Framework and the Test Tool. Next, attention will be given to standard tests and test libraries.

This section contains the following chapters:

- *Chapter 3, The Testability Framework*
- *Chapter 4, The Test Tools, Standard Tests, and Standard Test Libraries*

3
The Testability Framework

Having discussed the *why*, *when*, *what*, and a first, general *how* of test automation, it's time to turn our focus fully to Business Central. In this chapter, we will discuss how Business Central facilitates you to write and execute automated tests. The specific feature that enables this is called the **testability framework**.

The testability framework was introduced in the platform with Dynamics NAV 2009 Service Pack 1. For the first time, developers were able to build test scripts in C/AL code and run them in the client. At that time, however, you could only program **headless tests**; that is, tests that do not use the **user interface** (**UI**) to trigger the business logic. The testability framework was a follow-up to an internal tool called the **NAV Test Framework** (**NTF**) and had been used and worked on for a couple of years already.

NTF allowed tests to be programmed in C# and ran against the Dynamics NAV UI. It was a neat system, with a neat technical concept behind it. However, this *running tests against the UI* was one of the major reasons for leaving NTF behind. I seem to recall that it was the major reason because accessing business logic through the UI is slow – too slow. Too slow to allow the Microsoft Dynamics NAV development team to run all their tests against the various versions in a reasonable time.

Currently, Microsoft is supporting nine major versions (2017, NAV 2018, and Business Central 12, 13, 14, 15, 16, 17, 18) for 20+ countries, and each of these country versions is being built and tested at least once a day. Any delay in the tests has a huge impact on the build of these 8 versions.

In this chapter, we will have a look at what I call the **five pillars of the testability framework,** being the five technical features that make up this framework, as follows:

- Test codeunits and test functions
- The `asserterror` keyword
- Handler functions
- Test runner and test isolation
- Test pages

Technical requirements

The code examples can be found on GitHub at `https://github.com/ PacktPublishing/Automated-Testing-in-Microsoft-Dynamics-365- Business-Central-Second-Edition`.

The final code of this chapter can be found in `https://github.com/PacktPublishing/ Automated-Testing-in-Microsoft-Dynamics-365-Business-Central- Second-Edition/tree/main/Chapter 03`.

Details on how to use this repository and how to set up VS Code and the Business Central environment are discussed in *Appendix, Getting Up and Running with Business Central, VS Code, and the GitHub Project.*

The five pillars of the testability framework

In the following five sections, each *pillar* will be discussed and illustrated with a simple code example. Feel free to try them out yourself with or without the GitHub code examples. Being a hands-on book, we will get to a lot more relevant examples later on.

Pillar 1 – Test codeunits and test functions

Goal: Understand what test codeunits and test functions are and learn how to build and apply them.

The foremost important pillar of the testability framework is the concept of **test codeunits** and **test functions**.

Test codeunits

With test codeunits, we can create a suite of tests, with each test implemented as a so-called test function. The basics of a test codeunit are similar to a standard codeunit. You can set up its *OnRun* trigger, declare variables, and define functions in it. To make a codeunit a test codeunit, you need to set its Subtype to Test. Let's set up our first test codeunit:

```
codeunit 60000 MyFirstTestCodeunit
{
    Subtype = Test;
}
```

Having set the Subtype to Test makes this codeunit different from a standard codeunit in a couple of ways:

- It can contain test and handler functions next to the normal functions that we are used to when writing application code. Both will be described below.

- When executing a test codeunit, the platform will do the following:

    Run the OnRun trigger, if implemented, and each test function that resides in the test codeunit, from top to bottom.

    Report per test function if it succeeded, failed, or was skipped.

In the various examples, you will see and learn how to apply the test codeunit concept.

Test functions

As pointed out, a test codeunit can hold normal functions, like standard codeunits, and also so-called test functions. A Test function is defined by the MethodType tag [Test]:

```
[Test]
procedure MyFirstTestFunction()
begin

end;
```

A test function acts in many ways the same as a normal function, but differs in the following ways:

- It must be global.

- It cannot have arguments.

- It yields a result, which is either SUCCESS or FAILURE.

When SUCCESS is returned by a test, it means that no error occurred in the execution of the test. Consequently, when FAILURE is returned, the test execution did throw an error. This error could be due to various reasons, such as the following:

- Code execution hitting a TestField, FieldError, or Error call

- Data modifications not being fulfilled because of version conflicts, primary key conflicts, locks, or other runtime errors

The latter, a Test function returning FAILURE, brings us to another typicality of a test codeunit – when a test fails, the execution of a test codeunit doesn't halt. It continues to execute the next Test function. So, you know the result of all your tests, irrespective of a failure or success. Let's build two simple tests, one returning SUCCESS and the other FAILURE:

```
codeunit 60000 MyFirstTestCodeunit
{
    Subtype = Test;

    [Test]
    procedure MyFirstTestFunction()
    begin
        Message('MyFirstTestFunction');
    end;

    [Test]
    procedure MySecondTestFunction()
    begin
        Error('MySecondTestFunction');
    end;
}
```

Running this codeunit, the two test functions are executed from top to bottom. The message thrown by MyFirstTestFunction will show the following screenshot first:

Figure 3.1 – Message thrown by MyFirstTestFunction

After that, this message is shown, being a summary of the execution of the whole test codeunit:

Figure 3.2 – Result of MyFirstTestCodeunit

> **Note**
> The error did not appear like a message box but is collected by the platform and recorded as a part of the result of the failing test.

To be able to run the test codeunit, I built a simple page, MyTestsExecutor, with an action calling MyFirstTestCodeunit, executing both test functions in this test codeunit:

```
page 60000 MyTestsExecutor
{
   PageType = Card;
   ApplicationArea = All;
   UsageCategory = Tasks;
   Caption = 'My Test Executor';

   actions
   {
    area(Processing)
       {
         action(MyFirstTestCodeunit)
           {
             Caption = 'My First Test Codeunit';
             ToolTip = 'Executes My First Test Codeunit';
             ApplicationArea = All;
             Image = ExecuteBatch;
             RunObject = codeunit MyFirstTestCodeunit;
           }
       }
   }
}
```

If you are following me using the code on GitHub and having a hard time opening the `MyTestsExecutor` page, use any of the following options:

- Use *Alt + Q*, **Tell me what you want**, in the web client and search for `My Test Executor`.

- Set `startupObjectType` to `Page` and `startupObjectId` to `60000` in the `launch.json` of your VS Code project.

- Add `?page=60000` to the web client URL in the address bar of your browser, `http://localhost:8080/BC1630/?page=60000http://localhost:8080/66BC160/?page=60000`, when using an on-prem installation, or `http://localhost:8080/BC/?page=60000` when using a Docker installation.

- Launch the page directly from VS Code, making use of a VS Code AL extension such as **CRS AL Language Extension**.

> **Note**
>
> In *Chapter 4, The Test Tools, Standard Tests, and Standard Test Libraries*, we will discuss the Test Tool that has been conceived to run your tests. We did not use it here for two reasons:
>
> (1) The `MyTestsExecutor` page links directly to one of the traditional ways of executing a codeunit.
>
> (2) Using the Test Tool needs some introduction before we use it.

Pillar 2 – The asserterror keyword

Goal: Understand what the `asserterror` keyword means and learn how to apply it.

A substantial part of the business logic we implement specifies conditions under which a user action or a process should fail or stop to continue its execution. Testing the circumstances that lead to this failure is as important as testing the successful conclusion of an action or process.

The second pillar allows us to write tests that are focused on checking whether errors do occur; a so-called **positive-negative** or **rainy path** test. For example, posting errors out because a posting date has not been provided, or that, indeed, a negative line discount percentage cannot be entered on a sales line. To achieve this, the **asserterror** keyword should be applied in front of the `calling statement`:

```
asserterror <calling statement>
```

Let's use it in a new codeunit and run it:

```
codeunit 60001 MySecondTestCodeunit
{
  Subtype = Test;

  [Test]
  procedure MyNegativeTestFunction()
  begin
    Error('MyNegativeTestFunction');
  end;

  [Test]
  procedure MyPositiveNegativeTestFunction()
  begin
    asserterror Error('MyPositiveNegativeTestFunction');
  end;
}
```

The MyPositiveNegativeTestFunction function is reported as a SUCCESS, and, consequently, no error message is recorded:

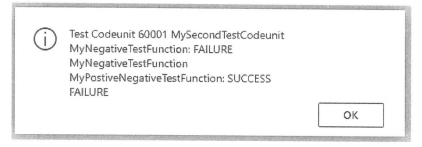

Figure 3.3 – Result of MySecondTestCodeunit

If the calling statement following the asserterror keyword throws an error, the system will continue executing the following statements in the test function. However, if the calling statement does not throw an error, the asserterror statement will cause one:

```
An error was expected inside an asserterror statement.
```

Where asserterror enables the test to continue with the next statement, it will not check the error as such. As we will see later, it is up to you to verify whether the expected error did occur or not. If there is no verification on the specific error following the asserterror, any error will make your test pass and it might be one that you were not expecting.

> **Note**
>
> If a successful positive-negative test does not report the error, this does not mean that the error did not occur. It is thrown, and, therefore, when a write transaction was performed, a rollback will happen. Any data *modifications*, done after a previous commit, will disappear.

Pillar 3 – Handler functions

Goal: Understand what handler functions are and learn how to build and apply them.

In our first test codeunit example, the `Message` statement results in the display of a message box. Unless we want to wait until a user presses the **OK** button, this message box stays there forever, halting the full execution of our test run. In order to be able to have a fully automated test run, we need a way to deal with any user interactions, such as a message box, a confirm dialog, a report request page, or a modal page.

For this, handler functions, also known as **UI handlers**, have been conceived. Handler functions are a special type of function that can only be created in test codeunits and aim at handling UI interactions that exist in the code under test. Handler functions enable us to fully automate tests without the need of a real user to interact with them. As soon as specific UI interactions occur, and a handler has been provided for it, the platform takes care of calling the handler as a substitute for real user interactions.

The `Test` function handler functions are defined by the `MethodType` tag. A number of the currently available values are shown in the following screenshot:

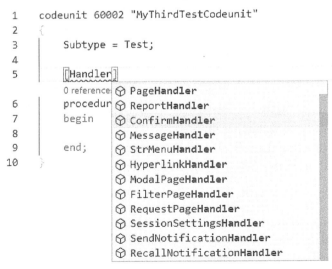

Figure 3.4 – A number of the currently available handler function types

Each handler function addresses a different type of user interaction object and needs different parameters to let it interact adequately with the platform. Let VS Code and the AL extension be your guide in finding the right signature for a handler function. The following screenshot shows you the signature of a `MessageHandler` when you hover over the function name:

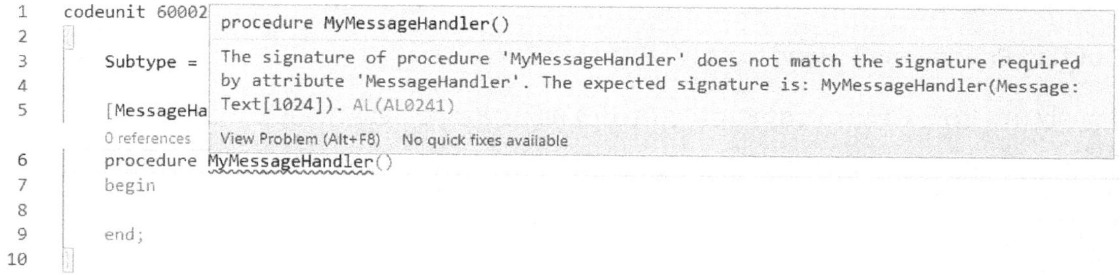

```
1    codeunit 60002    procedure MyMessageHandler()
2
3        Subtype =    The signature of procedure 'MyMessageHandler' does not match the signature required
4                     by attribute 'MessageHandler'. The expected signature is: MyMessageHandler(Message:
5        [MessageHa   Text[1024]). AL(AL0241)
                       0 references    View Problem (Alt+F8)    No quick fixes available
6        procedure MyMessageHandler()
7        begin
8
9        end;
10
```

Figure 3.5 – Syntax error on MessageHandler signature

In the case of a `MessageHandler` function, the signature is the text in the message box that would show to the user. Handing over this text to the `MessageHandler` enables you to determine if the right message was triggered.

> **Notes**
>
> (1) For a listing of the signature of each handler type, go to `https://docs.microsoft.com/en-us/dynamics-nav/how-to--create-handler-functions`.
>
> (2) The waldo's CRS AL Language Extension contains for each UI handler a code snippet with the right signature.
>
> (3) There is no handler for a drop-down control.

So, to get the `Message` statement handled automatically in our first test codeunit, we should create a `MessageHandler` function:

```
[MessageHandler]
procedure MyMessageHandler(Message: Text[1024])
begin

end;
```

But this is only half of the job, as this handler needs to be linked to the test that will execute the code calling `Message` one way or another. The `HandlerFunctions` tag is used to do this. Each handler function needs to be triggered in a `Test` function and must be added to the `HandlerFunctions` tag as text. If multiple handlers are needed, these will make up a comma-separated string, and only one handler function per handler type:

```
HandlerFunctions('Handler1[, Handler2,…] ')
```

Let's apply this to `MyFirstTestFunction` in a new codeunit and run it:

```
codeunit 60002 MyThirdTestCodeunit
{
  Subtype = Test;

  [Test]
  [HandlerFunctions('MyMessageHandler')]
  procedure MyFirstTestFunction()
  begin
    Message('MyFirstTestFunction');
   end;

  [MessageHandler]
  procedure MyMessageHandler(Message: Text[1024])
  begin

  end;
}
```

Instantly, rather than showing a message box first, the summary message of the execution of the whole test codeunit is shown:

Figure 3.6 – Result of MyThirdTestCodeunit

> **Notes**
>
> (1) Any handler function you add to the `HandlerFunctions` tag must be called at least one time in the `Test` function. If the handler is not called upon, because the user interaction it should handle doesn't happen, an error will be thrown by the platform, saying: **The following UI handlers were not executed**, listing the handlers not called upon.
>
> (2) The VS Code **AL Language** extension does not provide **Go to definition** on references of UI handlers. The **AL Code Actions** extension fixes this, so you might want to enable this extension in VS Code.

Pillar 4 – Test runner and test isolation

Goal: Understand what a test runner and its test isolation are and learn how to use and apply them.

Given the previous three pillars, we are positioned to write test cases as follows:

- Using **test codeunits** and **test functions**
- Either *sunny* or *rainy path*, the latter by applying the `asserterror` keyword
- With fully automated execution addressing any user interactions by applying **handler functions**

Do we need more?

As a matter of fact, yes, we do; we need a way to do the following:

- Run tests stored in multiple codeunits, control their execution, and collect and secure the results.
- Run tests in isolation, so that each (re)run of a test or test codeunit is done using the same data setup by rolling back any write transactions after concluding the test or test codeunit. This makes each test run **repeatable** and **deterministic**.

Both goals can be accomplished using a so-called **TestRunner** codeunit with certain test isolation. A test runner codeunit is defined by its `Subtype` and the isolation by its `TestIsolation`:

```
codeunit Id MyTestRunnerCodeunit
{
  Subtype = TestRunner;
  TestIsolation = Codeunit;
}
```

Below we discuss the test runner and its isolation in more detail.

> **Note**
> A lot of readers of the first edition of this book somehow got the idea that they needed to build a test runner themselves. Apparently, I did not make clear that there is no major reason to do that as Microsoft provides a number of tests runners. With pillar 4, I am just explaining the concept of a test runner.

Test runner

Like any other codeunit, a test runner codeunit can have an `OnRun` trigger and normal user-defined functions. With these, we can build an engine to get test codeunits executed, but we also need the ability to control the execution of test codeunits and their test functions and eventually collect their results. This is facilitated by two specific triggers called **OnBeforeTestRun** and **OnAfterTestRun** that can be added to a test runner.

When test codeunits are called from the `OnRun` trigger of a test runner, `OnBeforeTestRun` and `OnAfterTestRun` will be triggered by the system as follows:

- `OnBeforeTestRun` is triggered before the `OnRun` trigger of the test codeunit is executed, and again when each of its test functions is run.

- `OnAfterTestRun` is triggered after each test function has run and the test codeunit finishes.

Typically, the `OnBeforeTestRun` trigger is used to perform a test run pre-initialization. Next to that, it has a second agenda that is not clearly expressed in the function name. The framework is also using the outcome of the `OnBeforeTestRun` trigger to decide if a test codeunit should be skipped or not. Returning `TRUE`, the test codeunit or test function runs. Returning `FALSE`, it is skipped. This feature comes in handy when you temporarily want to deactivate a test codeunit.

Use the `OnAfterTestRun` trigger to perform post-processing, such as logging the result of each test. When the `OnAfterTestRun` trigger is run, the standard result message box of a test codeunit, as we have seen so far, is not shown.

Both `OnBeforeTestRun` and `OnAfterTestRun` are run in their own database transaction. This means that changes made to the database with each of these triggers are committed once their execution finishes. This is regardless of the test isolation set on the test runner.

> **Note**
>
> If you want to read more on the `OnBeforeTestRun` and `OnAfterTestRun` triggers, follow these links:
>
> **OnBeforeTestRun**: `https://docs.microsoft.com/en-us/dynamics365/business-central/dev-itpro/developer/triggers-auto/codeunit/devenv-onbeforetestrun-codeunit-trigger`
>
> **OnAfterTestRun**: `https://docs.microsoft.com/en-us/dynamics365/business-central/dev-itpro/developer/triggers-auto/codeunit/devenv-onaftertestrun-codeunit-trigger`

A test runner example

Usually, you do not have to create test runners yourself as they are provided with the AL Test Tool. We will discuss this in *Chapter 4*. However, if you would like to construct one, you might want to get inspiration from the standard test runner codeunit like `Test Runner - Isol. Codeunit (130450)`. Let's discuss step by step what this codeunit is intended to do (note that you can relate it to the code below):

1. A test runner codeunit like this will be called from the AL Test Tool – discussed in *Chapter 4* – and will get a reference handed over to the table behind the test tool. This is the `Test Method Line` table – see the `TableNo` property and the `Rec` variable used in the code.

2. In the `OnRun` trigger, the `Test Suite` record, related to the `Test Method Line`, is being retrieved. This is to verify that the test suite indeed exists and if so, to collect its data. Next, the current `Test Method Line` is safeguarded as we will need it later. Eventually, the `RunTests` method of the standard codeunit `TestRunnerMgt` is being called, which will run all tests that fall in the filters set on the `Rec` variable.

3. The `OnBeforeTestRun` trigger – its purpose discussed above – will get a number of parameters handed to it. It will pass them through to the `PlatformBeforeTestRun` method of the standard codeunit, `TestRunnerMgt`. This method will perform a number of preparatory actions like setting various values on the progress dialog and initiating a timer to allow for the calculation of the duration of a test.

4. The `OnAfterTestRun` trigger – its purpose discussed above – will update the same progress dialog, and update the `Test Method Line` with the final outcome of the test method related to this line:

```
codeunit 130450 "Test Runner - Isol. Codeunit"
{
  Subtype = TestRunner;
  TestIsolation = Codeunit;
  TableNo = "Test Method Line";
  Permissions =
    TableData "AL Test Suite" = rimd,
    TableData "Test Method Line" = rimd;

  trigger OnRun()
  begin
    ALTestSuite.Get(Rec."Test Suite");
    CurrentTestMethodLine.Copy(Rec);
    TestRunnerMgt.RunTests(Rec);
  end;

  var
    ALTestSuite: Record "AL Test Suite";
    CurrentTestMethodLine: Record "Test Method Line";
    TestRunnerMgt: Codeunit "Test Runner - Mgt";

  trigger OnBeforeTestRun(CodeunitId: Integer;
    CodeunitName: Text; FunctionName: Text;
    FunctionTestPermissions: TestPermissions): Boolean
  begin
    exit(
        TestRunnerMgt.PlatformBeforeTestRun(
          CodeunitId,
          CopyStr(CodeunitName, 1, 30),
          CopyStr(FunctionName, 1, 128),
          FunctionTestPermissions,
          ALTestSuite.Name,
          CurrentTestMethodLine.GetFilter("Line No."))
      );
  end;
```

```
    trigger OnAfterTestRun(CodeunitId: Integer;
      CodeunitName: Text; FunctionName: Text;
      FunctionTestPermissions: TestPermissions;
      IsSuccess: Boolean)
    begin
      TestRunnerMgt.PlatformAfterTestRun(
          CodeunitId, CopyStr(CodeunitName, 1, 30),
          CopyStr(FunctionName, 1, 128),
          FunctionTestPermissions,
          IsSuccess, ALTestSuite.Name,
          CurrentTestMethodLine.GetFilter("Line No.")
        );
    end;
  }
```

> **Note**
>
> You might wonder where to find the `Test Runner - Isol. Codeunit` (130450) we discussed just now. This codeunit resides in the Test Runner app. Accessing it in the symbols file of the Test Runner app will unfortunately not show the code. Instead, take the source code of the Test Runner app from the product DVD or the artifact you probably download when setting up your Docker container.

Test isolation

With a test runner, we can control the execution of all tests in one run, but we also need to have control over the data created in a test run. Each test codeunit or test function should run repeatably, creating exactly the same data over and over again. To achieve this, it must be possible to isolate one test run from the next. For this, the test runner `TestIsolation` property has been introduced with the following three options:

- `Disabled`: When selecting this value or not setting the `TestIsolation` property explicitly, as this is the default value, all database transactions are saved; after the execution of tests triggered by the test runner, the database will have changed compared to before running the test runner.

- `Codeunit`: When selecting this value, after a test codeunit execution has been completed fully, all data changes made to the database will be reverted/rolled back. Thus, data created in one test function is available in the next.

- `Function`: When selecting this value, when a single test function has been completed, all data changes made to the database will be reverted/rolled back. Thus, data created in one test function is **not** available in the next.

Related to this, it makes sense to share a couple of thoughts on running tests and their isolation:

- Test isolation applies to database transactions but does not apply to changes made outside of the database, to variables, including temporary tables, and the state of the client, like the work date.

- With test isolation, `Codeunit` or `Function`, **all** data changes will be rolled back, even if they were explicitly committed using the AL `Commit` statement.

- Running test codeunits outside of the test isolation of a test runner, or running them as a normal codeunit like we did through the `MyTestExecutor` page, will commit any database transaction.

- Using test isolation `Function` will give extra runtime overhead compared to `Codeunit`, resulting in longer execution time, as with the ending of each test function, the database changes have to be reverted.

- Setting the test isolation to `Function` might be unwanted as it fully disables dependencies between test functions. In the case of an extended test scenario, where intermediate results should be reported, this can be achieved by a series of individual, but interdependent, test functions. Using test isolation `Function` will not yield the right result.

- With the `TestIsolation` property of a test runner, we have control over how to revert data changes in a generic way; as we will later see, the test function `TransactionModel` tag allows us to have control of the transaction behavior of individual test functions.

When to use a specific test isolation?

Having three possible values for the `TestIsolation` property of the test runner, the question is: in what case you should be using one of them?

- Use `Disabled` when the data created should stick in the database, that is, not be rolled back when the test codeunit or function finishes.

  This could be wanted either when:

  (1) You want to be able to inspect the data resulting from the test, for example, when you start building tests and need to understand the data created.

  (2) Your tests do trigger other Business Central sessions like APIs. In that case, the data need to be committed in the database so the other session can pick it up.

- Use either `Codeunit` or `Function` when, after the tests are completed, the database should revert to each previous state, allowing you to rerun your tests or run the next collection of the test, in exactly the same circumstance.

  The `Codeunit` test isolation is the preferred one of the two as it gives lesser overhead, leading to faster execution.

> **Note**
>
> Microsoft does not promote the use of test isolation `Function`, and they are even considering deprecating the `Function` test isolation. This is reflected in the fact that they do not provide a test runner with this test isolation.
>
> There are two major reasons for this:
>
> (1) As mentioned earlier, using this test isolation will give extra overhead leading to longer execution time.
>
> (2) Using this test isolation, you need to make your tests independent of each other's data. While this is achievable, the challenge might be that the rollback works fine on the data but does not revert the state of the session the tests are run in.

Pillar 5 – Test pages

Goal: Understand what test pages are and learn how to apply them when testing the UI.

The initial trigger for adding the testability framework to the platform was to get away from testing the business logic through the UI. The testability framework enabled headless, and thus faster, testing of the business logic. And this is how the testability framework was implemented in NAV 2009 SP1. Pure headless testing. It included everything of the four pillars discussed so far, even though test isolation was implemented in a different way than it is today. It was previously not possible to test the UI.

Moving ahead, it became clear that sole headless tests excluded too much. How could we test business logic that typically resides on pages? For example, consider a product configurator in which options are displayed or hidden depending on values entered by the user. So, with NAV 2013, Microsoft added the fifth pillar to the testability framework: **test pages**.

A test page is a logical representation of a page and is strictly handled in memory displaying no UI. To define a test page, you need to declare a variable of the `TestPage` type:

```
PaymentTerms: TestPage "Payment Terms";
```

A `TestPage` variable can be based on any page that exists in the solution.

A test page allows you to mimic a user carrying out the following actions:

- Accessing the page
- Accessing its sub parts
- Reading and changing data on it
- Performing actions on it

You can achieve this by using the various methods that belong to the test page object. Let's build a small codeunit in which we use a couple of them:

```
codeunit 60003 MyFourthTestCodeunit
{
  Subtype = Test;

  [Test]
  procedure MyFirstTestPageTestFunction()
  var
    PaymentTerms: TestPage "Payment Terms";
  begin
    PaymentTerms.OpenView();
    PaymentTerms.Last();
    PaymentTerms.Code.AssertEquals('LUC');
    PaymentTerms.Close();
  end;

  [Test]
  procedure MySecondTestPageTestFunction()
  var
    PaymentTerms: TestPage "Payment Terms";
  begin
    PaymentTerms.OpenNew();
    PaymentTerms.Code.SetValue('LUC');
    PaymentTerms."Discount %".SetValue('56');
    PaymentTerms.Description.SetValue(
        PaymentTerms.Code.Value()
      );
    Error('Code: %1 \ Discount %: %2 \Description: %3',
        PaymentTerms.Code.Value(),
        PaymentTerms."Discount %".Value(),
        PaymentTerms.Description.Value()
      );
    PaymentTerms.Close();
  end;
}
```

Note that an error is forced to get some useful feedback on the summary message of the test codeunit. So, we get the following as a result:

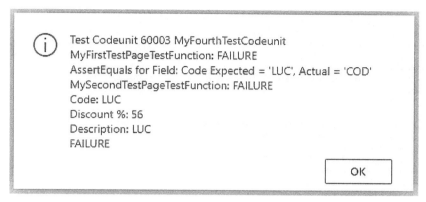

Figure 3.7 – Result of MyFourthTestCodeunit

> **Note**
>
> For a complete listing of all test page methods, you can access the following URLs:
>
> **TestPage**: https://docs.microsoft.com/en-us/dynamics365/
> business-central/dev-itpro/developer/methods-auto/
> testpage/testpage-data-type
>
> **TestField**: https://docs.microsoft.com/en-us/dynamics365/
> business-central/dev-itpro/developer/methods-auto/
> testfield/testfield-data-type
>
> **TestAction**: https://docs.microsoft.com/en-us/dynamics365/
> business-central/dev-itpro/developer/methods-auto/
> testaction/testaction-data-type

If you are running Microsoft 365 Business Central on-premises and you want to run tests using test pages, be sure that you have the **Page Testability** module installed. Creating a Docker container using *BCContainerHelper* will automatically install this:

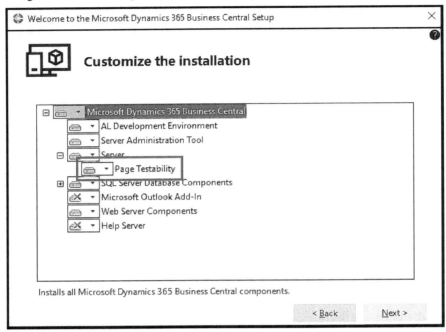

Figure 3.8 – Business Central Setup showing the Page Testability component

Summary

In this chapter, we discussed what the testability framework is by describing the five pillars it entails: the basic elements, test codeunits and test functions, the code keyword `asserterror`, handler functions to allow automatic handling of UI elements, the test runner to enable us to run tests in isolation, and, finally, test pages to build tests to check the behavior of pages.

In *Chapter 4*, you will learn about the test tool that resides in Business Central and enables you to run tests, and the set of standard tests being released by Microsoft with the product.

4

The Test Tools, Standard Tests, and Standard Test Libraries

With the previous chapter discussing the 5 pillars of the testability framework, the floor, as it were, was laid to start writing automated tests in Business Central. Ever since the introduction of the testability framework in NAV 2009 SP1, test automation has been a major part of Microsoft's work on Business Central. Not only getting automated tests in place, but also in providing a tool to run the tests and libraries with reusable helper functions to get test data set up. Using the testability framework, altogether Microsoft created a humongous set of automated tests to verify the standard application, a Test Tool feature to run the tests, and a vast number of test helper libraries to facilitate test data creation.

In this chapter, we will discuss the following topics in detail:

- Test tools
- Standard tests
- Test libraries

Technical requirements

Each of these components is provided by Microsoft as one or more extensions on the product DVD and in artifact feeds.

The Test Tool–being the C/AL Test Tool, see below–is also present in Business Central online, but you will only be able to fetch tests for the additional Microsoft extensions that reside in CRONUS, not for the base and system application. To get yourself a free trial, visit `https://dynamics.microsoft.com/en-us/dynamics-365-free-trial/`.

Details on how to set up your Business Central environment are discussed in *Appendix, Getting Up and Running with Business Central, VS Code, and the GitHub Project.*

Test Tools

Goal: Understand what the Test Tool entails and learn how to use and apply it.

The **Test Tool** is a feature that allows you to manage and run the automated tests that reside in the database that belong to either the standard application or any other extension and collect their results. With the various hands-on example tests, we will be using this tool a lot, and before we do this, let's elaborate a little bit on it.

You can easily access the Test Tool using the **Tell me what you want to do** feature in Dynamics 365 Business Central, as displayed in *Figure 4.1*.

Trying to access the Test Tool this way, you might notice Business Central can hold two test tools, as is also shown in the screenshot above. One called **Test Tool** and the other **AL Test Tool**. Depending on the setup of Business Central, it might even be the first one itself. When Microsoft built their first test tool, Business Central was still called Dynamics NAV and was fully C/AL and C/SIDE based; no extensions yet. Even though the technical objects that make up this tool were named *CAL Test <something>*, it is referenced in the UI as *Test Tool*. Moving forward, however, this test tool no longer covered the needs.

It was designed to be a tool that did more than just run the tests and return their results. Also, it has always been part of the Base Application and as such it most likely can be installed on every production environment where it does not belong; ever since Business Central 2019 wave 1, called BC 15 for short, C/SIDE is gone, and Business Central is fully extension based.

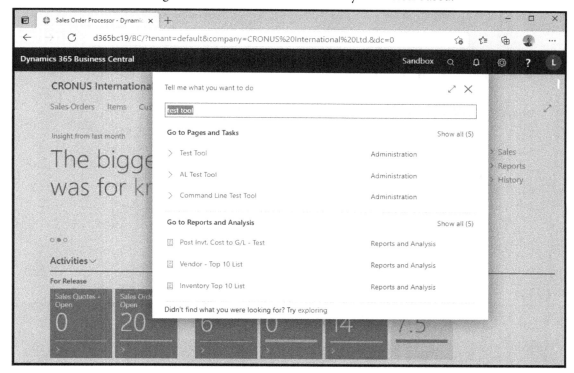

Figure 4.1 – Using "Tell me what you want to do" feature

With this, Business Central development moved towards a more modern way of working using **continuous integration** and **continuous delivery (CI/CD)** pipelines; a leaner version of the test tool was obvious. This new version was constructed in a separate extension and was baptized **AL Test Tool**. The *old* test tool–which we will, from now on, refer to as **C/AL Test Tool**–will therefore in due time be deprecated. In this book, for this very reason, we will only discuss and use the AL Test Tool.

If the extension holding the **AL Test Tool** wasn't installed, when using the **Tell me what you want to do** feature to search for the test tool, the **AL Test Tool** will not be listed.

Notes

(1) In *Chapter 9, How to Integrate Test Automation in Daily Development Practice*, CI/CD pipelines in the context of test automation will be discussed.

(2) How to get the AL Test Tool in your installation will be described in *Appendix, Getting Up and Running with Business Central, VS Code, and the GitHub project*.

(3) A description of **CAL Test Tool** can be found on my blog: `https://www.fluxxus.nl/index.php/BC/how-to-use-the-old-test-tool-aka-cal-test-tool`.

Now, selecting **AL Test Tool** from the **Tell me what you want to do** result list opens the **AL Test Tool** page, as shown in *Figure 4.2*:

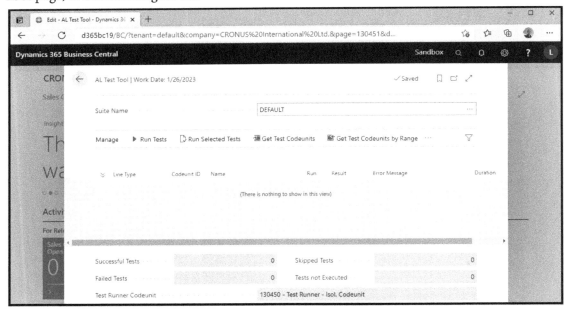

Figure 4.2 – AL Test Tool with the DEFAULT test suite

Typically, this is what you will see when in a clean Business Central installation, or in a company where the **AL Test Tool** has not been used yet: a worksheet page displaying an empty test suite called **DEFAULT**. You populate a test suite by adding one or more test codeunits that reside in the database. And once your test suite contains them you can run the tests. In the next sections, we will explain how to add and run test codeunits and discuss the result of a test run.

Adding tests to a test suite

To add tests to a test suite you use either the **Get Test Codeunits** or **Get Test Codeunits by Range** action. We will show next the steps to take using either of these actions to select test codeunits to be added to a test suite.

Adding tests using the Get Test Codeunits action

Take the following steps using the **Get Test Codeunits** action to include codeunits to a test suite:

1. Select the **Get Test Codeunits** action to open the **Select Tests** page.

   Unsurprisingly, it shows the test codeunits we created in *Chapter 3*, *The Testability Framework*, possibly followed one or more standard test codeunits depending on the setup of your Business Central environment.

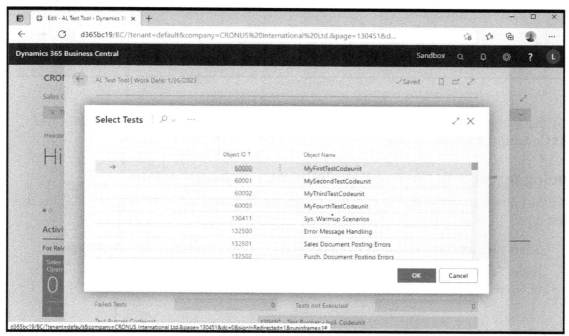

Figure 4.3 – Using Get Test Codeunits to select test code units

2. Select the four test codeunits from **60000** through **60003**.

3. Click **OK**.

The suite now shows, for each test codeunit, a line with **Line Type** = **Codeunit** and, linked to this line and indented, all its test functions (**Line Type** = **Function**), as shown in *Figure 4.4*:

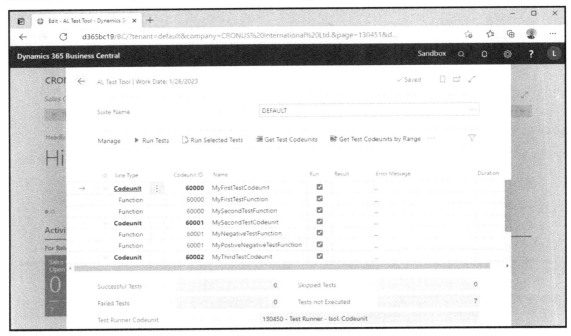

Figure 4.4 – AL Test Tool populated

Now that you know how to add tests to a test suite, let's look at how to add tests using the **Get Test Codeunits by Range** action.

Adding tests using the Get Test Codeunits by Range action

Take the following steps using the **Get Test Codeunits by Range** action to include codeunits to a test suite:

1. Select the **Get Test Codeunits by Range** action to open the **Select Tests by Range** page.

2. In the **Selection Filter**, fill in the ID range from which you want the test codeunits to be selected:

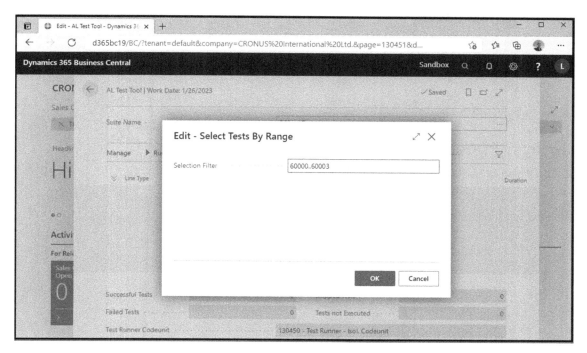

Figure 4.5 – Using Get Test Codeunits by Range to select test code units

3. Click **OK**.

The suite now will show a screen similar to the steps followed for the **Get Test Codeunits** action, in which also for each test codeunit a line with **Line Type** = **Codeunit** and, linked to this line and indented, all its test functions (**Line Type** = **Function**) are inserted.

> **Note**
>
> You use either action, **Get Test Codeunits** or **Get Test Codeunits by Range**, to include any test codeunit in a test suite. This can be either adding new test codeunits to a new, empty test suite or appending them to an existing one. Appending an already present test codeunit will create a new entry for it in the test suite.

In this section, you learned how to populate a test suite using the **DEFAULT** test suite that already resides in Business Central. You will learn how to create a new test suite next, when we discuss the standard tests.

Running the tests

Once you have added your tests to the test tool, you can run them. For this, we have two actions at our disposal:

- **Run Tests**
- **Run Selected Tests**

Running tests using the Run Tests action

The **Run Tests** action allows you to run the tests that go with the selected, or active, line, or all tests.

If the selected line is a *function* line the following dialog pops up:

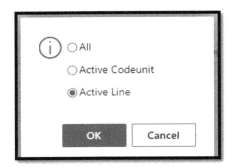

Figure 4.6 – Run Test dialog for function line

If you select **All** before you click **OK**, all tests will be executed. But with **Active Codeunit**, only the tests of the codeunit on the selected line, or the codeunit that holds the active function line, will be executed. And finally, with **Active Line**, only the test function your cursor is on will be executed.

If the selected line is a *codeunit* line the following dialog pops up:

Figure 4.7 – Run Test dialog for codeunit line

Now, given the suite is already set up–see *Figure 4.4*–let's go step by step to run the tests:

4. First, select the **Run Tests** action.

5. On the dialog that opens, select **All**.

6. Press **OK** to get the test run.

Now, all four test codeunits will be run and each test will yield a result, **Success** or **Failure**. Also, note the **Duration** column displaying the time it took to execute a test codeunit/function:

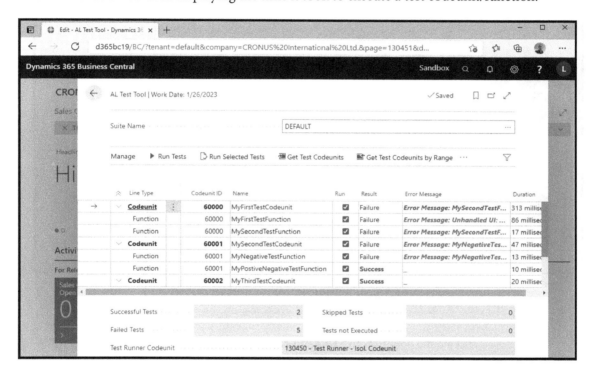

Figure 4.8 – AL Test Tool with executed tests

For each failure, the **Error Message** field will display the error that caused the failure and the **Call Stack** of the test execution. As you can see, the **Error Message** field has a hyperlink. If you click on it the full message, the test error and call stack is shown in a dialog, allowing you to read it easily. *Figure 4.9* illustrates an example:

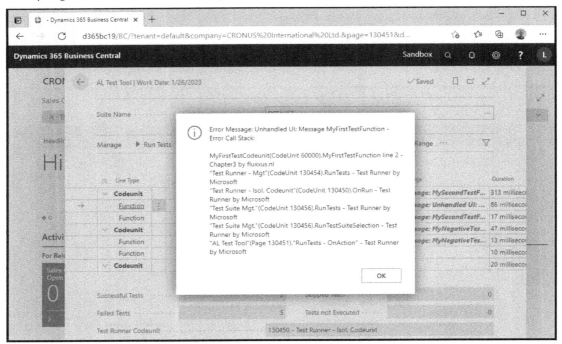

`Figure 4.9 – AL Test Tool with error message dialog

Notes

(1) Had we selected the option **Active Codeunit**, only the selected codeunit would have been executed.

(2) Although running `MyFirstTestFunction` in *Chapter 3* did not throw an error *Figure 4.9* clearly displays it does when run in the test tool. The reason for this is that running tests by means of a test runner will throw an error when UI is not handled automatically.

Running tests using the Run Selected Tests action

The **Run Selected Tests** action allows you to run only those tests that go with the lines you have selected. This could be single or multiple function or codeunit lines. Once you click on the action, the tests are executed instantly, that is, no dialog pops up. This action helps you to efficiently (re) run a test you are working on with no dialogs interfering, thus saving you some time.

Selecting a test runner

When you open the **AL Test Tool** for the first time, **Test Runner - Isol. Codeunit** (130450) is selected, as is shown at the bottom of *Figure 4.2*. As the name indicates, this is a test runner with test isolation set to Codeunit, which is the one you will probably use most. We will elaborate on this in a number of contexts later in the book.

Running the test by selecting either the **Run Tests** or **Run Selected Tests** action will call the test runner codeunit to get the job done:

- Preparing the test run
- Running the tests
- Collecting the results

If you need a different test runner, follow these steps:

1. On the test tool, select the **Select Test Runner** action.
2. On the **Select TestRunner** page, select the test runner of your choice.
3. Click **OK** to effectuate your choice.

Depending on the size of your web browser window, the **Select Test Runner** action might be shown or be hiding underneath the three dotted action (**Show the rest**).

Any remaining action on the test tool will be discussed later in the book as it makes more sense to discuss them when they are needed. In the next section, we will talk about the standard tests.

Standard tests

Goal: Get to know the basics of the standard tests provided by Microsoft.

Ever since NAV 2016, Microsoft has made its own application test collateral a part of the product. A humongous set of tests was initially delivered as a `.fob` file on the product DVD. Since Business Central is fully extension based, these tests, aka **standard tests**, are delivered in over 30 extensions. These standard test apps mainly contain test codeunits, but there are also a number of supporting tables, pages, reports, and XMLport objects in them.

For Dynamics 365 Business Central, the whole set contains over 36,000 tests in over 1,400 test codeunits, for w1 and local functionality for each country in which Microsoft releases. With every bug that's fixed, and with every new feature introduced in the application, the number of tests grows. The standard tests have been built over the last ten-plus years, and they cover all functional areas of Business Central.

Let's set up a new suite in the Test Tool called **ALL W1** now:

1. Open the **AL Test Tool**.
2. Click on the **Assist Edit** button in the **Suite Name** control.
3. In the **AL Test Suites** pop-up window, select **New**.
4. Populate the **Name** and **Description** fields.
5. Click **OK** to return to the **AL Test Tool**.

The newly created test suite will open:

Figure 4.10 – Select the newly created test suite ALL W1

Now, using the **Get Test Codeunits by Range** action and setting the **Selection Filter** to 1..200000, let Business Central fetch all test codeunits, as shown in *Figure 4.11*:

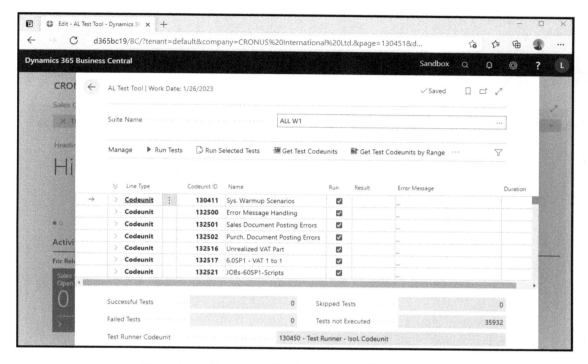

Figure 4.11 – Test suite ALL W1 containing all standard tests

Be aware that it might take a while to get all the standard tests loaded in the **AL Test Tool**. Also, running all standard tests in one run will take hours to complete with quite a substantial number of errors. I will elaborate on both in *Chapter 10, Getting Business Central Standard Tests Working on Your Code*.

Reading the names of these test codeunits will give you a rough impression of what they entail.

Standard tests are delivered in over 30 extensions. Some go with the system and Base Application, while others relate to separate Microsoft extensions. Discussing all these in detail is outside the scope of this book. We will, however, take a somewhat closer look at the extension **Base Application**.

Notes

(1) I did remove the test codeunit we created in *Chapter 3, The Testability Framework*, before I made the screenshot for *Figure 4.11* to only have the standard tests in view.

(2) The various test extensions can be either found on the product DVD or artifacts feeds in the `Test` subfolder of each extension folder in the `Applications` folder. The test apps that go with the Base Application can be found in `Applications\BaseApp\Test`.

Base Application tests

The **Base Application** contains most major functionality, such as Finance, Sales, Purchase, Inventory, Service Management, Production, Warehouse Management, Jobs, Resources, and HR:

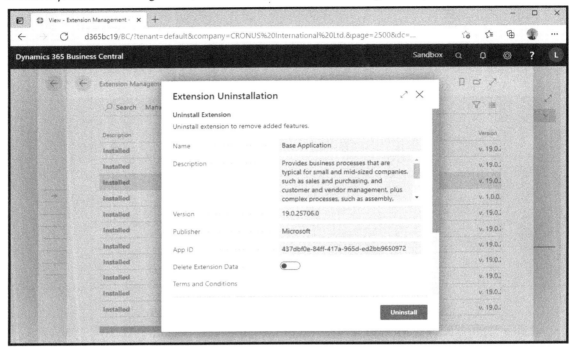

Figure 4.12 – The Base Application extension

`Test apps

The **Base Application** is accompanied by 34 test apps. These apps are listed in *Figure 4.13*:

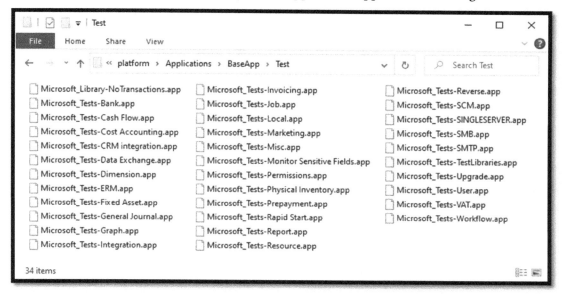

Figure 4.13 – The Base Application test extensions

Let's pick out a few of them to highlight some details.

Tests-SCM (Supply Chain Management)

The SCM test app **Tests-SCM** is the largest one by far and contains the largest set of tests, around 9,400 tests in over 230 codeunits, testing warehouse and production functionality.

Tests-ERM (Enterprise Resource Management)

Even though it's the second largest, the ERM test app, **Tests-ERM**, is probably the most important one as its tests cover the kernel of Business Central: G/L, sales, purchase, and inventory. It altogether consists of somewhat more than 300 test codeunits with over 9,000 tests.

Tests-Misc (Miscellaneous)

Test-Misc is, with about 3,150 tests in over 160 test codeunits, the third-largest test app that goes with the **Base Application**. It contains a wide variety of tests covering, among others, Company Initialization, Data Classification, Data Migration, Email Merge/Attachments, Change Log, Job Queue, and Notifications.

Tests-Workflow

With approx. 1,000 tests in about 60 test codeunits, the **Tests-Workflow** test app is the fourth-largest test app belonging to the **Base Application** and tests all kinds of workflow-related scenarios.

> **Note**
> The total number of tests residing in a test app was collected by installing the test app and populating a test suite with the test codeunits. The total number of test codeunits for a test app was drawn from the codeunit source files in the test app.

Unit Tests (UTs)

Most of the test apps contain functional, end-to-end tests. But there are also codeunits that hold **unit tests** (**UTs**). These are marked by the addition of UT to their name. Some examples are as follows:

- `Codeunit 134155 - ERM Table Fields UT`
- `Codeunit 134164 - Company Init Unit Test II`
- `Codeunit 134825 - UT Customer Table`

> **Note**
> Want to read more on unit and functional tests?
> Go to: `https://www.softwaretestinghelp.com/the-difference-between-unit-integration-and-functional-testing/`.

UI tests

With headless testing being the initial trigger for bringing the testability framework into the platform, it's no surprise that the clear majority of standard test codeunits comprises headless tests. Test codeunits that are meant to test the **user interface** (**UI**) are marked using `UI` in their name. Some examples are as follows:

- `Codeunit 134280 - Simple Data Exchange UI UT`
- `Codeunit 134339 - UI Workflow Factboxes`
- `Codeunit 134927 - ERM Budget UI`

Note, however, that these are not the only test codeunits addressing the UI. Any other codeunits might contain one or more UI tests, where, in general, the bulk will be headless tests.

Reports

As I am often asked how to test reports, it is worth mentioning, as a last category, those test codeunits that are dedicated to testing reports. Search for any test codeunit that is marked with the word `Report` in its name. You will find several dozens of them. The following are a couple of examples:

- Codeunit 134063 - ERM Intrastat Reports
- Codeunit 136311 - Job Reports II
- Codeunit 137351 - SCM Inventory Reports - IV

The **Base Application** test app **Tests-Report** contains almost 800 more tests that check many things related to reports, although some of the other apps also have report-related tests. The test functions, however, do not always contain the word `Report`.

> **Note**
>
> In *Chapter 8, From Customer Wish to Test Automation – the TDD Way*, we will elaborate on how to test a report.

Categorization by FEATURE

By the inspection above, we got an impression of what kind of tests this Base Application test collateral is made up of. Microsoft, however, has a better structured categorization, which so far, due to low priority, hasn't been explicitly shared with the outside world. Now that automated testing is being picked up more and more, pressure is mounting on Microsoft to put this on higher priority. But for now, we can access it already inside most of the test codeunits. You need to look for the FEATURE tag. This tag is part of the **Acceptance Test-Driven Development** (**ATDD**) test case design pattern, which we will be discussing later in *Chapter 5, Test Plan and Test Design*. Using the FEATURE tag, Microsoft categorizes their test codeunits and, in some cases, individual test functions. Note that this tagging is far from complete as not all test codeunits have it, yet.

Have a look at the (partial) abstract of the following codeunits.

Codeunit 134000 – ERM Apply Sales/Receivables

- OnRun:

  [FEATURE] [Sales]

- [Test] procedure VerifyAmountApplToExtDocNoWhenSetValue:

  [FEATURE] [Application] [Cash Receipt]

- [Test]procedure PmtJnlApplToInvWithNoDimDiscountAndDefDimErr:

    [FEATURE] [Dimension] [Payment Discount]

Codeunit 134012 – ERM Reminder Apply Unapply

- OnRun:

    [FEATURE] [ERM] [Reminder] [Sales]

- [Test]procedure CustomerLedgerEntryFactboxReminderPage:

    [FEATURE] [UI]

In later chapters, we will look in more detail at various standard test functions. You will see how to take them as examples for your writing own tests (*Chapter 5*, *Chapter 6*, *Chapter 7*, and *Chapter 8*), and how to get them to run on your own solution (*Chapter 10*).

Now it's time to have a look at our last topic: test libraries.

Note

At this very moment, the standard test suite objects are to be found in the following ID ranges:

(1) 132000 to 139999: w1 tests

(2) 144000 to 149999: local tests

Standard libraries

Goal: Get to know the basics about the standard test helper libraries provided by Microsoft.

Supporting their Base Application tests, Microsoft has created a nice and very useful collection of helper functions in approx. 100 library codeunits. These helper functions are gathered in the **Test-TestLibraries** test app and range from random data and master data generation to standard generic and more specific check routines.

Need a new item? You might make use of the CreateItem or CreateItemWithoutVAT helper functions in Library - Inventory (codeunit 132201).

Need some random text? Use the AlphabeticText helper function in Any (codeunit 130500).

Want to get the same formatted error messages when verifying your test outcome? Use one of the helper functions in the `Library Assert` (codeunit 130002), such as `IsTrue`, `AreNotEqual`, and `ExpectedError`.

> **Note**
>
> When referencing their helper functions in the standard libraries from your extension, you will need to define a dependency on that extension in the `app.json` of your extension.
>
> See *Appendix, Getting up Up and Running with Business Central, VS Code, and the GitHub project*, for more details on this.

Finding useful standard helper functions

A frequently reappearing question during my workshops is:

How do I know if these libraries contain a helper function that I need in my own test? Is there an overview of the various helper functions?

Unfortunately, there is no overview of all available helper functions for Dynamics 365 Business Central. However, up to NAV 2018, a `.chm` help file containing this information was included in the `TestToolKit` folder on the product DVD.

You might want to make use of this, but I always use a very simple method: a quick file search in a folder where I have stored all standard test objects. Let's say I need a helper that will create for me a service item; I might open VS Code on that folder and search for `CreateServiceOrder`, as shown in *Figure 4.14*:

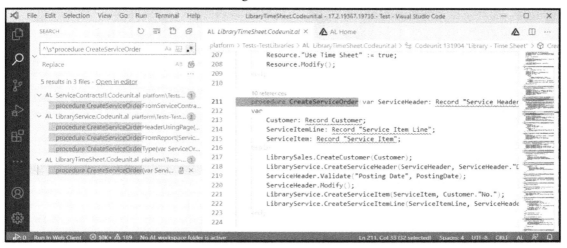

Figure 4.14 – Using VS Code file search

A while ago I did write a blog post about various ways of, and tips on, finding a helper function in the Microsoft test libraries that might fulfill my need: *How-to: Find and Use standard test Helper Functions*. Be welcome to make use of it. I really want to encourage you to read this post. You can find it here: `https://www.fluxxus.nl/index.php/BC/how-to-find-and-use-standard-test-helper-functions`.

In *Section 3*, *Designing and Building Automated Tests for Microsoft Dynamics 365 Business Central*, when building tests, we will rely on my searching methods to find various standard helper functions, making our tests efficient and consistent.

Some generic and very useful standard libraries

The majority of the approx. 100 library codeunits relate to very specific application features, such as `Library - Sales` (130509) or `Library - Marketing` (131900). To discuss all of them is far out of the reach of this book. There is, however, an interesting, small number of generic libraries that fully make sense to have a closer look at:

- `Library Assert` (130002)
- `Any` (130500)
- `Library - Utility` (131000)

Library Assert

`Library Assert` (130002) holds a consistent set of helper functions to be used when verifying the result of an automated test. Basically, any verification will check an *expected* (set of) value(s) against an *actual* (set of) value(s). It is not rocket science to write code to do that. By using one of the helper functions from `Library Assert` we ensure that you will get standardized error messages if the expected and actual values are not equal. Over time, when reading the error of any failing test, you will be able to recognize easily what kind of error occurred, given that your verification helper functions are making use of `Library Assert`.

This is a complete listing of all helper functions in `Library Assert`:

- `AreEqual(Expected: Variant; Actual: Variant; Msg: Text)`
- `AreEqualDateTime(Expected: DateTime; Actual: DateTime; Msg: Text)`
- `AreNearlyEqual(Expected: Decimal; Actual: Decimal; Delta: Decimal; Msg: Text)`

- `AreNotEqual(Expected: Variant; Actual: Variant; Msg: Text)`
- `AreNotNearlyEqual(Expected: Decimal; Actual: Decimal; Delta: Decimal; Msg: Text)`
- `AssertNoFilter()`
- `AssertNothingInsideFilter()`
- `AssertPrimRecordNotFound()`
- `AssertRecordAlreadyExists()`
- `AssertRecordNotFound()`
- `ExpectedError(Expected: Text)`
- `ExpectedErrorCode(Expected: Text)`
- `ExpectedMessage(Expected: Text; Actual: Text)`
- `Fail(Msg: Text)`
- `IsFalse(Condition: Boolean; Msg: Text)`
- `IsTrue(Condition: Boolean; Msg: Text)`
- `KnownFailure(Expected: Text; WorkItemNo: Integer)`
- `RecordCount(RecVariant: Variant; ExpectedCount: Integer)`
- `RecordIsEmpty(RecVariant: Variant)`
- `RecordIsNotEmpty(RecVariant: Variant)`
- `TableIsEmpty(TableNo: Integer)`
- `TableIsNotEmpty(TableNo: Integer)`

We will illustrate the purpose and use of various of them in *Section 3, Designing and Building Automated Tests for Microsoft Dynamics 365 Business Central*, when building tests.

> **Note**
>
> `Library Assert` is a replacement for the `Assert` (130000) library codeunit as Microsoft is slowly refactoring all their tests and libraries. The "old" tests and libraries, including `Assert`, are to be found in the `Test` folder of the `BaseApp` folder (`\platform\Applications\BaseApp`) on the product DVD and artifacts feed. The newer ones will be stored in `testframework` folder (`\platform\ Applications\testframework`).

Any (130500)

The `Any` library is a replacement of the old `Library – Random` (130440) and provides a number of helper functions allowing to create pseudo-random values for simple data types, such as integer, decimal, and text. These are all the helper functions that `Any` contains:

- `AlphabeticText(Length: Integer): Text`

- `AlphanumericText(Length: Integer): Text`

- `DateInRange(MaxNumberOfDays: Integer): Date`

  with *overloads* `DateInRange(StartingDate: Date; MaxNumberOfDays: Integer): Date`

  `DateInRange(StartingDate: Date; MinNumberOfDays: Integer; MaxNumberOfDays: Integer): Date`

- `DecimalInRange(MaxValue: Integer; DecimalPlaces: Integer): Decimal`

  with *overloads* `DecimalInRange(MinValue: Decimal; MaxValue: Decimal; DecimalPlaces: Integer): Decimal`

  `DecimalInRange(MinValue: Integer; MaxValue: Integer; DecimalPlaces: Integer): Decimal`

- `Email(): Text`

  with *overload* `Email(LocalPartLength: Integer; DomainLength: Integer): Text`

- `GuidValue(): Guid`

- `IntegerInRange(MaxValue: Integer): Integer`

  with *overload* `IntegerInRange(MinValue: Integer; MaxValue: Integer): Integer`

- `UnicodeText(Length: Integer) String: Text`

Except for the `GuidValue()` helper function, all other functions create data based on the AL system function `Random()`.

Along with the functions listed above, `Any` also features a `SetSeed()` function that lets you set the seed of the `Random()` system function.

> **Note**
>
> Pseudo-random means that every time you call one of the helper functions of Any in the same sequence of code execution, the provided value will be the same, unless you have changed the seed of the Random() functions by means of SetSeed().
>
> With automated tests, this is exactly what you want as your tests should be repeatable. Using "Any" of these helper functions provides your test a value which exact value is not relevant.

Library – Utilities (131000)

Library – Utilities contains, next to generic functions to, for example, convert data, different kinds of helper functions closely related to the platform. These helper functions let you generate and get number series, find the minimum or maximum value of a field, check the existence of a field in a table, get the value of a specific property of a field, compare records, and more. It's unfortunately too extensive to list here all the 50+ helper functions that reside in Library – Utilities.

On top of those generic platform helper functions, Library – Utilities also holds a number of (pseudo) random data generators, most of which have become redundant with Any.

Summary

In this chapter, we discussed in a nutshell what the Test Tool is, and you learned how to use it to run your tests, or even run the collection of tests Microsoft has built and provided us with. You got a short overview of the various categories of tests this vast collection contains. You concluded this chapter with a brief description of the *100* libraries containing useful helper functions to support your own test writing.

Now that you have understood the various test features that exist in Dynamics 365 Business Central, you are ready to go and start designing and coding your own tests. We will start *Chapter 5, Test Plan and Test Design*, to introduce several design patterns that facilitate easier fand consistent test case coding.

Section 3: Designing and Building Automated Tests for Microsoft Dynamics 365 Business Central

With this section, you have reached the centerpiece of this book, in which you will learn how to plan, design, and build automated tests. We will discuss the various concepts and design patterns. By making use of this and the features and tools discussed in the previous section, you will deep dive into the implementation of automated tests. In this section, you will move from the basics of headless and UI testing to more advanced techniques, and how to handle a positive-negative test.

This section contains the following chapters:

5
Test Plan and Test Design

Having had a look at the testability framework, the Test Tool, and the standard tests and libraries, I have shown you what is available in the platform and the application that will allow you to create and execute automated tests. And that's what we are going to do in this part of this book. But let's step back and not just bluntly dive into creating code. Firstly, I would like to introduce a couple of concepts and design patterns that will allow you to conceive your tests more effectively and efficiently. At the same time, these concepts will make your tests more than just a technical exercise and will help you to get your entire team involved. After all, testing is a team effort.

I am surely not going to bother you with formal test documentation, and the top-down approach with eight defined stages, from test plan, through test design/case specification to test summary report. That's way out of the scope of this book and, not in the least, beyond the daily practices of most Dynamics 365 Business Central implementations. Nevertheless, a couple of thoughts on plan and design spent upfront will give you leverage in your work.

The following topics will be covered in this chapter:

- No plan, no test
- Setting up a test plan
- Test case design patterns

- Test data setup design patterns
- Using customer wish as test plan
- And what about unit and functional tests?

If I have been overloading you with information and you want to get your hands on coding, you might want to jump to *Chapter 6, From Customer Wish to Test Automation – the Basics* . There, we will start exercising all the things discussed so far and later in this chapter. However, if you find out you are missing background details, come back to this chapter and get yourself informed.

No plan, no test

Goal: Understand why tests should be planned, and designed, before they are coded and executed.

I guess I am not far off when saying that most of the application testing done in our world falls under the term of **exploratory testing** or **ad-hoc testing**. That is: testing done manually by experienced persons that know the application under test and have a good understanding and feeling of how to *break the thing*. But this is most often exercised with no explicit design and no reproducible, shareable, and reusable scripts. In this world, we typically don't want developers to test their own code as they, consciously or unconsciously, know how to use the software and evade issues. Their mindset is *how to make it* (work), not *how to break it*.

With automated tests, it will be developers that will code them. And more often than not, it will be the same developers that do the application coding. So, they need a design of what tests to code. Tests that will cover a broad set of scenarios. Tests that are *sunny* and *rainy*, that is, tests succeeding without exception versus tests that verify the exceptions. *Headless* and *UI tests*, being tests triggered by code only versus tests triggered through the UI.

A full-fledged test plan would describe the various kinds of tests that should be executed, such as **performance, application**, and **security tests**; the conditions under which they must be performed; and the criteria that make them successful. Our test plan will only address application tests, as that is the focus of this book: how to create application test automation.

In my humble opinion, the **test plan** and **test design**, like any other deliverable, are owned by the team. It is a joint effort to agree upon the two; an agreement between the product owner, tester, developer, functional consultant, and key user. And if there is *no plan*, there will be *no test*. Test plan and test design are objects that help the team discuss their test effort, reveal the holes in their thoughts, and let it mature while working on it. In addition, as will be discussed later, it is a possible way of putting your requirements down, permitting your team to efficiently get from requirements to test and application code.

Ideally a test plan should be made up and agreed upon before any coding is done; as such it should be part of the requirements, and might, as part of agile practices, be iterated. We will discuss this in more extent in the *Using customer wish as test plan* section. Most of us, however, need to get test automation in place for features already built. Our challenge is how to get a test plan set up; raising, among others, the following questions:

- Where to start?

- What should it contain?

- When is it sufficient, that is, when does it hold a complete enough set of test cases?

Let's step into the next section and expand on this.

> **Note**
> If you want to learn more about formal test documentation, this Wikipedia article could be a first stepping stone: `https://en.wikipedia.org/wiki/Software_test_documentation`.

Setting up a test plan

Goal: Learn how to set up a test plan for an existing feature.

A test plan should list all possible tests that are needed to verify whether the application, or just a singular feature, is doing what it is expected to do. First of all, it should entail both the **sunny scenarios**, a.k.a. **positive tests** and often defined as user stories or use cases, and **rainy scenarios**, a.k.a. **positive-negative tests**, that is, tests that check that exceptions are handled. Next to that, it should clearly discriminate between tests that should be automated and those that do not have to. Regarding the first ones – those that need to be automated – it needs to be pointed out whether they check the business logic without accessing the UI, a.k.a. **headless tests**, and those that do need to verify UI specifics (**UI tests**). And finally, it has to prioritize tests, enabling the team to focus successfully on what needs to be tested beyond any doubt, and thus managing the risks of testing the wrong things as in the end, there might be time constraints at play, urging the team to make choices. With a test plan, we construct the *test list* we talked about in *Chapter 2, Test Automation and Test-Driven Development*.

Once the test plan holds a complete, or complete enough set of test cases, they need to be detailed out in a test design as will be discussed next in *Test case design patterns*. But before we do, how could you efficiently and effectively set up a test plan? Where to start?

Break down your application or feature

Depending on the complexity of the application or feature, setting up a test plan might be more or less challenging. But in all cases, the following procedure has proven to be of great help:

1. Given the application or feature, determine at the highest level into what **functional areas** it can be split up.

2. Determine for each **area** whether you are able to describe the behavior that should be tested.

3. If so, identify all the tests that check that behavior. If not, split up the area you're working on into **sub areas** and go to *step 2* and repeat the same for each **sub area** until each is covered by a complete (enough) list of tests.

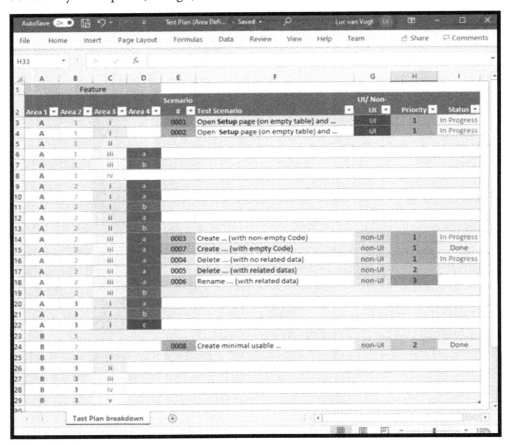

Figure 5.1 – Example of a work in progress test plan breakdown

To perform this exercise I typically will use a spreadsheet, like in Excel, (see *Figure 5.1*) to keep an overview, to easily add extra columns, and eventually end up with a list of tests that can easily be transferred to some other format or tool and, not in the least, allow it to be reviewed by the team leading to an improved list. And not to forget to carry out the next step: designing each test. For that, let's move to the next topic.

Notes

(1) Even though the goal of this section stated "learn how to set up a test plan for an existing feature," you can also apply this breakdown procedure if you are still developing your application code before putting any focus on testing.

(2) The example test pan can be found on the GitHub repo in `https://github.com/PacktPublishing/Automated-Testing-in-Microsoft-Dynamics-365-Business-Central-Second-Edition/tree/main/Excel sheets/Examples/Chapter 05`.

Test case design patterns

Goal: Learn the basic patterns for designing tests.

Once a **test plan** is in place, it is time to detail the various tests it defines. In this section, we move from a **test plan** into a **test design**.

If you have been testing software, you might know that each test has a similar overall structure. Before you can perform the action under test, for example, the posting of a document, the data needs to be *set up*. Then, the action will be *exercised*. And finally, the result of the action has to be *verified*. In some cases, a fourth phase applies, a so-called **teardown**. This is used to revert the system under test to its previous state before a next test can take off, making each test repeatable.

Four-phase testing

The four phases of a test case design pattern are listed as follows:

1. Setup
2. Exercise
3. Verify
4. Teardown

This **four-phase test** design pattern was used by Microsoft in the early years of C/SIDE test coding. Like the following test function example, taken from codeunit `SCM Inventory Misc. III` (137295), you will encounter it in a vast number of older test codeunits:

```
[Test]
[HandlerFunctions('SalesInvoiceStatisticsPageHandler,
    CreditMemoConfirmHandlerYes')]
[Scope('OnPrem')]
procedure PstdSalesInvStatisticsWithSalesPrice()
var
  SalesLine: Record "Sales Line";
  PostedSalesInvoice: TestPage "Posted Sales Invoice";
  DocumentNo: Code[20];
begin
  // Verify Amount on Posted Sales Invoice Statistics after
  // posting Sales Order.

  // Setup: Create Sales Order, define Sales Price on
  // Customer,.
  Initialize;
  CreateSalesOrderWithSalesPriceOnCustomer(SalesLine, WorkDate);
  LibraryVariableStorage.Enqueue(SalesLine."Line Amount");
      // Enqueue for SalesInvoiceStatisticsPageHandler.

  // Exercise: Post Sales Order.
  DocumentNo := PostSalesDocument(SalesLine, true);
      // TRUE for Invoice.

  // verify: Verify Amount on Posted Sales Invoice Statistics.
  // Verification done in SalesInvoiceStatisticsPageHandler
  PostedSalesInvoice.OpenView;
  PostedSalesInvoice.Filter.SetFilter("No.", DocumentNo);
  PostedSalesInvoice.Statistics.Invoke;
end;
```

Nowadays, however, Microsoft uses the **Acceptance Test-Driven Development (ATDD)** design pattern, which we will discuss next.

> **Note**
>
> For a short and clear description of the four-phase design pattern, please refer to the following link:
>
> `http://xunitpatterns.com/Four Phase Test.html`

Acceptance Test-Driven Development

Like the **four-phase test** design pattern, **ATDD** addresses the setup, exercise, and verify stages in a test. But it introduces a more complete structure, which is also closer to the user as tests are described from the user's perspective. The pattern is defined by the following so-called tags:

- **FEATURE**: Defines what feature the test or collection of test cases is testing.

- **SCENARIO**: Defines for a single test the scenario being tested.

- **GIVEN**: Defines what data setup is needed; a test case can have multiple **GIVEN** tags when data setup is more complex.

- **WHEN**: Defines the action under test; each test case should have only one **WHEN** tag.

- **THEN**: Defines the result of the action, or more specifically the verification of the result; if multiple results apply, multiple **THEN** tags will be needed.

You might notice that the **ATDD** design pattern has no equivalent of the **teardown** phase in the four-phase test design patterns. As mentioned, ATDD is user-oriented, and the teardown is a more technical exercise and is, in most cases, handled by the Business Central platform, like the test isolation of a test runner codeunit (see *Chapter 3, The Testability Framework*). But of course, if needed, a teardown section should be coded at the end of a test, although very often we do not need to as some parts are automatically torn down by means of the Test Isolation that was defined in the Test Runner codeunit.

> **Note**
>
> On request of readers, we have assigned specific colors to the different tags:
>
> **Purple** for FEATURE
>
> **Gold** for SCENARIO
>
> **Red** for GIVEN
>
> **Green** for WHEN
>
> **Blue** for THEN

The following test example, taken from test codeunit ERM Bank Reconciliation (134141), displays an **ATDD** design pattern-based test:

```
[Test]
[Scope('OnPrem')]
procedure VerifyDimSetIDOfCustLedgerEntryAfterPosting
  BankAccReconLine()
var
  BankAccReconciliation: Record "Bank Acc. Reconciliation";
  BankAccReconciliationLine: Record
    "Bank Acc. Reconciliation Line";
  StatementAmount: Decimal;
  CustomerNo: Code[20];
  CustLedgerEntryNo: Integer;
  DimSetID: Integer;
begin
  // [FEATURE] [Customer]
  // [SCENARIO 169462] "Dimension set ID" of Cust. Ledger
  //                    Entry should be equal "Dimension Set ID" of
  //                    Bank Acc. Reconcilation Line after posting
  Initialize;

  // [GIVEN] Posted sales invoice for a customer
  CreateAndPostSalesInvoice(CustomerNo, CustLedgerEntryNo,
    StatementAmount);

  // [GIVEN] Default dimension for the customer
  CreateDefaultDimension(CustomerNo, DATABASE::Customer);

  // [GIVEN] Bank Acc. Reconcilation Line with
  //         "Dimension Set ID" = "X" and
  //         "Account No." = the customer
  CreateApplyBankAccReconcilationLine(
    BankAccReconciliation, BankAccReconciliationLine,
    BankAccReconciliationLine."Account Type"::Customer,
    CustomerNo, StatementAmount, LibraryERM.CreateBankAccountNo);
```

```
DimSetID :=
  ApplyBankAccReconcilationLine(
    BankAccReconciliationLine,
    CustLedgerEntryNo,
    BankAccReconciliationLine."Account Type"::Customer, '');

  // [WHEN] Post Bank Acc. Reconcilation Line
  LibraryERM.PostBankAccReconciliation(BankAccReconciliation);

  // [THEN] "Cust. Ledger Entry"."Dimension Set ID" = "X"
  VerifyCustLedgerEntry(
    CustomerNo, BankAccReconciliation."Statement No.", DimSetID);
end;
```

Before this test was coded, however, a test case design was conceived to be handed off to the developer to enable the test coding:

```
[FEATURE] [Customer]
[SCENARIO 169462] "Dimension set ID" of Cust. Ledger Entry should
                  be equal "Dimension Set ID" of
                  Bank Acc. Reconcilation Line after posting
[GIVEN] Posted sales invoice for a customer
[GIVEN] Default dimension for the customer
[GIVEN] Bank Acc. Reconcilation Line with "Dimension Set
        ID" ="X" and "Account No." = the customer
[WHEN] Post Bank Acc. Reconcilation Line
[THEN] "Cust. Ledger Entry"."Dimension Set ID" = "X"
```

This **ATDD** closely relates to **TDD (Test-Driven Development)**, where a test always precedes the creation of related application code. **ATDD** tests, however, always relate to the behavior the end-user is requesting.

> **Note**
>
> The Scope attribute in the test example from test codeunit ERM Bank Reconciliation (134141) is set to OnPrem, which ensures that the test can only be run in an on-prem Business Central environment and not in a SaaS environment.

A note on test verification

In my workshops, it is mainly developers that participate. For them, when working on test automation, one of the hurdles to cross is the verification part. It makes perfect sense to all of them that data setup, the `GIVEN` part in ATDD, has to be accounted for, not to mention the action under test in the `WHEN` part. The `THEN` part, the verification, however, is something easily neglected, especially if it is a developer's assignment to come up with the `GIVEN`-`WHEN`-`THEN` design. Some might ask: why should I need verification if the code executes successfully? The answer is quite simple: non-erroring tests do not ensure that the behavior is the right behavior. Your tests always need to check the outcome. Things like:

- Is the data created, the right data, that is, the expected data?
- Is the error thrown, in case of a positive-negative test, the expected error?
- Is the confirm that was handled indeed the expected confirm?

Sufficient verification will make sure your tests will stand the test of time. You might want to put the next phrase as a poster on the wall:

A test without verification is no test at all!

Feel free to add a bunch of extra exclamation marks.

Applying test case design patterns successfully

The ultimate goal of a **test case design pattern** is to support you in defining effective and efficient tests. Tests that verify the behavior of a feature as simple, as fast, and as complete as possible. Tests that are also easy to code – again: effectively and efficiently. Each test case should, therefore, contain only the bare information needed to get this done. If one `GIVEN` is enough, do not provide more. If a three-word sentence captures the `SCENARIO` well, stick to it. At the same time, be sure to use enough words to fully seize the test case.

If a customer is needed in a scenario and all its attributes are of no importance, the following `GIVEN` suffices:

```
[GIVEN] Customer
```

But if the customer is a foreign customer, the `GIVEN` should be:

```
[GIVEN] Foreign customer
```

As a team, you need to find out and learn what is *enough*; you need to develop a common language to enable you to apply test case design patterns successfully. Keep it simple and yet thorough. Hopefully, you will find out that scenarios very often can be achieved in a less complicated way than projected at first. Like application code, tests and test code are developed in an iterative way. We will surely run into this working through the examples further on.

> **Note**
>
> For more information on **ATDD**, you can follow these links: `https://en.wikipedia.org/wiki/Acceptance_test-driven_development` or `https://docs.microsoft.com/en-us/dynamics365/business-central/dev-itpro/developer/devenv-extension-advanced-example-test#describingyour-tests`.

Test data setup design patterns

Goal: Learn the basic patterns for setting up test data.

When you carry out your manual tests, you know that most of the time is consumed by setting up the right data. Being a real IT pro, you will think up ways of doing this as efficaciously as possible. You might have thought of making sure of the following:

- A basic setup of data is available, which will be the foundation for all of the tests you are going to execute.

- For each feature under test, additional test data exists upfront.

- Test-specific data will be created on the fly.

This way, you have created yourself a number of patterns that help you efficiently get your test data setup done. These are what we call **test data setup** design patterns, or **(test) fixture** design patterns, and each has its own name:

- The first one is what we call a **prebuilt fixture**. This is test data that is created before any of the tests are run. In the context of Dynamics 365 Business Central, this could be a prepared database, such as CRONUS, the demo company that Microsoft provides.

- The second pattern is known as **shared fixture**, or lazy setup. This concerns the setup of data shared by a group of tests. In our Dynamics 365 Business Central context, this concerns generic master data, supplemental data, and setup data, such as customer and currency data and a rounding precision, all needed to run a group of tests.

- The third and last pattern is **fresh fixture**, or fresh setup. This entails data particularly needed for a single test, such as an empty location, a specific sales price, or a document to be posted.

When automating tests, we will make use of these patterns for the following reasons:

- **Efficient test execution**: Even though an automated test seems to run at the speed of light, building up a test collateral over the years will increase the total execution time, which might easily run into hours. The shorter the automated test run will be, the more it will be used.

- **Effective data setup**: When designing test cases, it is straight away clear what data will be needed and at what stage; this will speed up the coding of the tests.

Whatever data setup design patterns you will use, note that the trigger to applying either of them should always be the `GIVEN` section in your scenario. Or in other words: try, really try, to ensure that your scenario always relates to the data setup it specifically needs. With this I am not saying you should always detail the requested data setup with `GIVEN` clauses describing all atomic parts of the fixture. You could, for example, start this way:

```
[GIVEN] Company setup A
```

This could entail a very specific company setup that everyone knows about and which is captured in one fixture creation method.

> **Note**
>
> Read more on fixture patterns on the xUnit patterns initiative web site: `http://xunitpatterns.com/Fixture Setup Patterns.html`.
>
> Note that there is much more to formalize in the test data setup. In our test coding in the next chapters, we will utilize a couple more of the patterns mentioned.

Test fixture, data agnostics, and prebuilt fixture

As stated in the introductory chapter, automated tests are *reproducible*, *fast*, and *objective*. They are reproducible in their execution as the code is always the same. But this does not guarantee whether the outcome is reproducible. If, each time a test is run, the input to the test is different, then presumably, the output of the test might also be different. The following three things help to ensure your tests are always reproducible:

- Make a test run on the same fixture.

- Make a test follow the same code execution path.

- Make a test verify the outcome based on the same and sufficient set of criteria.

To have full control of the **fixture**, it is highly preferable to let your automated tests create the data they need anew, with each run. In other words, do not rely on the data present in the system before tests are run. Automated tests should be agnostic of any data residing in the system under test. Consequently, running your tests in Dynamics 365 Business Central should not rely on the data present in the database, be it CRONUS, your own demo data, or customer-specific data.

Yes, you might need customer-specific data to reproduce a reported issue, but once the issue is fixed and the test automation is updated, it should be possible to run data agnostically. Because of this, as a starting point, we will not use the **prebuilt fixture** pattern in any of our tests. You could, however, construct reusable shared fixture generators to be combined into one process to be triggered before you run all your tests, manually or automated. In this way, you can save the time not having this data to be created anew by every test codeunit. We will touch this topic later again.

If you've ever run the standard tests, you might have noticed that quite a number of them are not **data agnostic**. They highly rely on the data present in CRONUS. You might also have noticed that this applies to the older tests. At present, standard tests strive to be data agnostic.

Note

I often end up in a discussion on not using **prebuilt fixture**.

Why could my tests not make use of data already present in the database? This makes the designing and coding of the tests much more efficient, right?

Fully true, but this is short-term thinking. In the long run you will suffer from it, like Microsoft does with their older tests that rely on CRONUS data. What happens if the "already present" data changes? Do your tests still run successfully? Chances are high they won't and you need to update the GIVEN part of your tests. This might sound simple, but often you have to get hold of the context again and find all the additional *nuts and bolts*.

Test fixture and test isolation

To start each set of tests with the same fixture, we will take advantage of the test isolation feature of the test runner codeunit, as discussed in *Chapter 3* in the *Pillar 4 - test runner and test isolation* section. Using the test isolation value Codeunit of a test runner and putting a coherent set of tests in one test codeunit, a generic *teardown* for the whole test codeunit is set. It will assure that, at the termination of each test codeunit, the fixture is reverted to its initial state. If the test runner utilizes the **Function** test isolation, it would add a generic teardown to each test function, which in turn also adds time to the test execution.

Shared fixture implementation

You might have observed in the two Microsoft test functions, used as examples for the four-phase and ATDD patterns, that each test starts with a call to a function named `Initialize`, right after the scenario description. `Initialize` contains the standard implementation of the shared fixture pattern (next to a generic fresh fixture pattern we will elaborate on later) as follows:

```
local procedure Initialize()
begin
  // Generic Fresh Setup
  LibraryTestInitialize.OnTestInitialize(<codeunit id>);

  <generic fresh data initialization>

  // Lazy Setup
  if isInitialized then
    exit();

  LibraryTestInitialize.OnBeforeTestSuiteInitialize(
    <codeunit id>);

  <shared data initialization>

  isInitialized := true;
  Commit();

  LibraryTestInitialize.OnAfterTestSuiteInitialize(
    <codeunit id>);
end;
```

As indicated above, when talking about **shared fixture**, `Initialize` should be used to create generic data that is needed to run a group of tests. Things like **master data**, **supplemental data**, and **setup data**. And of course, this data should preferably have been defined by at least one GIVEN tag.

When calling `Initialize` at the start of each test function in the same test codeunit, the lazy setup part will just be executed once, since only with the first call to `Initialize` will the Boolean variable be `false`. Each test function has a call to `Initialize` because you can run an individual test function from the AL Test Tool, and you want to make sure that every test has the data that it needs. Note that `Initialize` also incorporates three hooks, that is, event publishers, to allow extending `Initialize` by linking subscriber functions to it:

- `OnTestInitialize`
- `OnBeforeTestSuiteInitialize`
- `OnAfterTestSuiteInitialize`

In *Chapter 10, Getting Business Central Standard Tests Working on Your Code*, we will specifically make use of these publishers. The lazy setup part of `Initialize` is what xUnit patterns calls **SuiteFixture** setup.

Fresh fixture implementation

A **fresh fixture** can be (partly) set up in a generic way as per the implementation in the `Initialize` function as discussed previously. This is for data that needs to be created or cleaned out at the start of each test. A fresh setup specifically needed for one test only is what is to be found inline in the test function that implements the test, matching its related GIVEN tag(s).

The generic fresh setup part of `Initialize` is what xUnit patterns call **Implicit Setup**. The test-specific fresh setup is called **Inline setup**.

Applying test data setup design patterns successfully

As with the test case design patterns (see *Applying test case design patterns successfully*), your team needs to find out and learn what fixture applies best to the tests you're working on. The same adage applies: keep it simple and yet thorough and use what works for your team.

Using customer wish as test plan

Goal: Learn why and how to describe **the customer wish** in the form of a test plan and test design, and as such make it easier to test automation.

A past ideal of development was a staged one, where each phase would finish before the next started, just like a waterfall, with water flowing from one level to the other. Moving from the customer wish through requirements gathering to analysis, to design, to coding, to testing, and finally operation and maintenance. Each phase would have its deadlines, and documented deliverables are handed off to the next phase. One of the major drawbacks of this system is its lack of responsiveness to changing insights, resulting in changing requirements. Another is the significant overhead of documents produced.

In the recent decade or two, agile methodologies have become a general practice for tackling these drawbacks. And thus, introducing with a test plan an extra document – and with it a test design – to your development practice is maybe not what you have been waiting for, even though I can guarantee that your development practices will get leveraged.

What if your test plan/design could be a kind of unified document? Be the input for each discipline in your project? The same truth shared at each level? What if you could *kill five birds with one stone*? If you could transform your customer wish into requirements in one format as input for all implementation tasks? Let's see how this can be done.

It's a common practice to transform your customer wish into requirements using user stories or use cases. But in my humble opinion, these often entail two major omissions:

- They tend to only define the *sunny path* and have no explicit description of the *rainy scenarios*. How should your feature behave under non-typical input? How should it error out? As mentioned before, this is where a tester's mind deviates from a developer's mind: how to make it break versus how to make it work. This would definitely be part of a test plan. So, why not promote the test plan, and ultimately the test design, to become the requirements, that is the detailed customer wish, describing the wanted behavior? Or outside-in: write your requirements like a test design using the ATDD pattern.

- They are most often conceptual descriptions of a customer wish. Due to this, each role involved with the implementation of this wish makes its own interpretation when casting this into a **behavior** the application code is to enact, and test to verify. As we know, I guess, with each interpretation things tend to be left out, leading in the least to unintended behavior, or worse, and most probably, to buggy code.

Both omissions can be overcome with the following steps:

1. Break down the **customer wish** into a list of tests – ATDD scenarios! – describing how the software should behave, making it our primary vehicle of communication in the next steps.

2. Implement this behavior with **application code**.

3. Execute **manual tests** methodically, to check the behavior.

4. Code **test automation** to check the behavior.

5. Update the **documentation** on how your solution behaves.

Or, in case of a **TDD** approach where you code your test first:

1. Break down the **customer wish** into a list of tests – ATDD scenarios! – describing how the software should behave, making it our primary vehicle of communication in the next steps.

2. Code **test automation** to check the behavior.

3. Implement this behavior with **application code** and run the automated tests.

4. Update the **documentation** on how your solution behaves.

This is what I am advocating in presentations, workshops, and training, and this is what implementation partners are picking up. With this, your test automation will be a logical result of previous work. New insights, resulting in requirement updates, will be reflected in this list and accordingly in your test automation. Whereas your current requirement documentation might not always be in sync with the latest version of the implementation, they will be when promoting your test design to requirements, as your automated tests will have to reflect the latest version of your app code. In this manner, your test automation is your up-to-date documentation. Killing five birds with one stone, indeed; getting your team *facing the same direction.*

As we will be doing in the next chapters, we specify our requirements as a test plan, at the feature and scenario level, using the FEATURE and SCENARIO tags. Then, following this, the test design will be shaped by providing details to the scenarios using the GIVEN, WHEN, and THEN tags. Have a peek preview of how this looks in the following example, which is one scenario for the LookupValue extension that we are going to work on in the next chapters:

```
[FEATURE] LookupValue UT Sales Document
[SCENARIO #0006] Assign lookup value on sales quote
                 document page
[GIVEN] Lookup value
[GIVEN] Sales quote document page
[WHEN] Set lookup value on sales quote document
[THEN] Sales quote has lookup value code field populate
```

Moving with Business Central from an on-premises world with low update frequency to a cloud world with high update frequency, we can no longer afford to get things lost in interpretations. We can no longer afford to lose as much time as we used to in the sequence from customer wish through requirement specifications to testing. Even more, we cannot bear as many bugs in the product as we used to anymore once it's *out there*. That's why we need to get everyone involved, *facing the same direction*, by transforming requirements from a conceptual description eventually into a behavioral description. This is what ATDD and the like are doing as ultimately, software implements behavior and nothing more.

> **Note**
> Want more inspiration on getting your team *facing the same direction*? Go and watch my Areopa webinar, *My requirements specification – all facing the same direction*, on YouTube: `https://www.youtube.com/watch?v=TchUOb76uRs`.

And what about unit and functional tests?

Except for a short note in *Chapter 1, Introduction to Automated Testing*, and mentioning a couple of MS test codeunits that apparently hold unit tests, we haven't paid any attention to the concept of unit tests so far. In this chapter, however, discussing the need for a test plan entailing all tests verifying the behavior of the feature, it makes a lot of sense to pick up *that gauntlet* laying in sight: what about **unit tests**?

My short answer to this question is: please, go ahead and implement them.

But quite obviously that might not be the answer you're looking for. So, let me work this out a little bit more in a number of bullet points:

- In my humble opinion, unit tests are the sole responsibility of developers. They build them to check the validity of the *atomic units that altogether make up a feature*. These **atomic units** are the procedures and methods they have created to implement a feature. As such, unit tests are not part of the requirements.

- Testing the atomic unit's unit tests do not verify the feature as such, just its components that make up the feature.

- To test the feature, we rise to the level of **functional testing**: to verify whether the feature behaves as requested.

Now how do unit tests relate to **functional tests**?

- It is well possible that unit tests show that the atomic units and the components are perfectly fine, while it appears the feature is not. Therefore, you cannot do without functional tests.

- This *you-cannot-do-without-functional-tests* and the fact that the Business Central world hasn't had a lot of experience with unit testing, I tend to have us mainly, or first, focus on functional testing. That's the testing we are used to. That's the testing that ultimately should prove the requested feature.

And then the *please, go ahead and implement them* comes around the corner again. Meaning: with all arguments given, and taken action upon, if you find value in them, indeed go ahead and implement unit tests.

> **Note**
> We'll discuss more on unit testing in *Chapter 12, Writing Testable Code*.

Test case design pattern and unit tests

My primary testing focus being a functional one, using the ATDD pattern for test case design makes perfect sense. You might, however, wonder what pattern to use for **unit tests**. As any test is about checking (*verification*), you can use any pattern in the outcome of an action (*exercise*) performed on a given environment (*setup*), be it the four-phase or ATDD pattern. Or, as often used in the unit tests context, **AAA (Arrange, Act, Assert)**.

As all patterns are telling the same, my preference would be to use ATDD for all tests.

Functional testing versus integration testing

Where functional testing was used in the previous section, the concept of **integration testing** could also have been used. That I choose to only use **functional** testing is that for Business Central, in many cases, the two could be intertwined. Often with functional testing in Business Central, the tests stretch over multiple features and functional modules, implicitly testing the integration of these. Of this I am aware, and I am doing short on integration testing as this also entails integration between pure technical components.

Summary

Test automation will profit from a structured approach and, for this, we introduced the test plan and test design, test patterns, and test fixture patterns. Next, we discussed how the step from requirements to their implementation could be made efficient and effective, including related test automation. This was done by shaping the requirement specifications as a test plan and getting your team *facing the same direction*.

Now, in the next chapter, *Chapter 6*, we will utilize these patterns to implement test code. You might have been waiting long, but we do finally get into coding automated tests.

Further reading

Having worked in this world of Business Central development for over two decades, I did run into quite some methodologies and read about them. In the context of testing and test automation, for a long time my focus was explicitly on these topics per se. In the last couple of years, trying to get my team, and other teams too, *facing the same direction*, my focus turned more and more from app and test code implementation to the requirements, as you might have sensed above. In this turn someone pointed me to a book worthwhile reading, lifting the need for detailed test descriptions from the, surely *my*, context of test automation. If you can spare some time and money, you can go and read this book:

- Gojko Adzic, *Specification by Example – How successful teams deliver the right software*, Manning, 2011.

6
From Customer Wish to Test Automation – the Basics

You are technically fully set to start writing tests at this point. This is because you know how the testability framework functions, you know the test toolkit, you know about the existence of the standard test libraries, and you have been provided with various patterns to allow you to design efficient and effective tests.

But what are you going to test? What's your business case? What are the customer wishes you are going to implement? In this chapter, you will start applying the principles and techniques discussed in the previous chapters and will build a number of basic automated tests.

As such, this chapter covers the following topics:

- From customer wish to test automation
- Test example 1 – a first headless test
- Test example 2 – a first positive-negative test
- Test example 3 – a first UI test
- Headless versus UI

Technical requirements

The **LookupValue** extension we are going to build in this book can be found on GitHub: `https://github.com/PacktPublishing/Automated-Testing-in-Microsoft-Dynamics-365-Business-Central-Second-Edition`.

This repository also includes an Excel file containing a list of all ATDD scenarios that apply to the LookupValue extension. You can find the various ATDD sheets here: `https://github.com/PacktPublishing/Automated-Testing-in-Microsoft-Dynamics-365-Business-Central-Second-Edition/tree/main/Excel sheets`. Even though we will pick out specific scenarios as examples to elaborate on, note that the whole list of scenarios has been conceived upfront describing, in full, the customer wish.

The final code of this chapter can be found in `https://github.com/PacktPublishing/Automated-Testing-in-Microsoft-Dynamics-365-Business-Central-Second-Edition/tree/main/Chapter 06 (LookupValue Extension)`.

Details on how to use this repository and how to set up VS Code and the Business Central environment are discussed in *Appendix, Getting Up and Running with Business Central, VS Code, and the GitHub Project*.

From customer wish to test automation

Our customer wishes to extend standard Dynamics 365 Business Central with an elementary feature: the addition to the `Customer` table of a **lookup field** to be populated by the user. This field has to be carried over to the whole bunch of sales documents and also needs to be included in the warehouse shipping.

Data model

Even though the purpose of such a field will be very specific, we will generically name it `Lookup Value Code`. As with any other lookup field in Business Central, this `Lookup Value Code` field will have a table relation (foreign key) with another table, in our case, a new table called `Lookup Value`.

Figure 6.1 schematically describes the data model of this new feature, with the new table in the middle and the extended standard tables on the left and right sides:

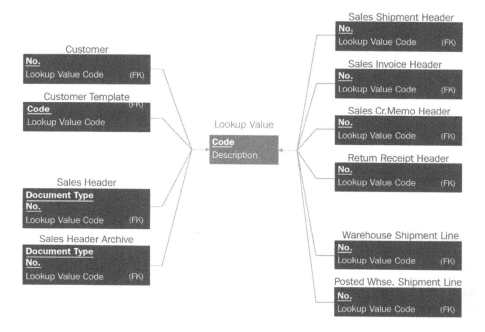

Figure 6.1 – Lookup Value data model relational diagram

The `Lookup Value Code` field has to be editable on all tables except for the posted document header tables, that is, `Sales Shipment Header`, `Sales Invoice Header`, `Sales Cr.Memo Header`, `Return Receipt Header`, and `Posted Whse. Shipment Line`.

Business logic

In compliance with standard Business Central behavior, the following business logic applies:

- When creating a customer from a customer template, the `Lookup Value Code` field should be inherited from `Customer Template` to `Customer`.

- When selecting a customer in the `Sell-to Customer` field on a sales document, the `Lookup Value Code` field should be inherited from the `Customer` to `Sales Header`.

- The `Lookup Value Code` field on a sales document should be able to be set or changed manually by the user.

- When posting a sales document, it is mandatory that the `Lookup Value Code` field is populated.

- When posting a sales document, the `Lookup Value Code` field should be inherited from `Sales Header` to the header of the posted document. That is, `Sales Shipment Header`, `Sales Invoice Header`, `Sales Cr.Memo Header`, or `Return Receipt Header`.

- When archiving a sales document, the `Lookup Value Code` field should be inherited from `Sales Header` to `Sales Header Archive`.

- When restoring an archived sales document, the `Lookup Value Code` field should be restored from `Sales Header Archive` to `Sales Header`.

- When creating a warehouse shipment from a sales order, the `Lookup Value Code` field should be inherited from `Sales Header` to `Warehouse Shipment Line`.

- The `Lookup Value Code` field on `Warehouse Shipment Line` should be able to be set or changed manually by the user.

- When posting a warehouse shipment, the `Lookup Value Code` field should be inherited from `Warehouse Shipment Line` to `Posted Whse. Shipment Line`.

- When copying a sales document using the **Copy Document** feature, the `Lookup Value Code` field should be inherited from source `Sales Header` to target `Sales Header`.

- Extend the `Lookup Value Code` field to sales document reports.

Based on these requirements, the **LookupValue** extension will be built as discussed next, including automated tests.

> **Note**
>
> Our first steps into test automation do relate to the daily practice of most of us: having the application code already in place and then working on the tests. As such, this chapter, and the next, *Chapter 7, From Customer Wish to Test Automation – Next Level*, will not go the TDD way yet. This we will pick up in *Chapter 8, From Customer Wish to Test Automation – the TDD Way*.
>
> You can find application-ready code in the `src` folder of the VS Code project on GitHub that goes with each chapter.

Converting our customer wish into a test plan and test design

Now that you have a clearly defined customer wish, you need to get everyone *facing the same direction* while setting up our **test plan** first. Then, using the **ATDD** test case pattern with the FEATURE, SCENARIO, GIVEN, WHEN, and THEN tags, we will detail out the test plan to become our **test design**.

Setting up the test plan

Let's practice the breakdown method as discussed in *Chapter 5, Test Plan and Test Design*, to find the basic functional parts that make up our customer wish. Our overall *container*, our main **area** is **LookupValue**. From an ATDD perspective, we could point this out to be our FEATURE. As this is, in my humble opinion, too big, we need to split this up into a next **area** level. We will call this the **sub feature** level. So, what sub features can we come up with?

- Customer
- Sales Document
- Customer Template
- Warehouse Shipment
- Sales Archive
- Inheritance
- Posting
- Report
- Permissions
- Copy Document

Let me shortly explain the rationale behind each of these areas.

Customer

Next to the `Lookup Value` entity, the **Customer** forms the basis of the **LookupValue** feature. Here is where a lookup value is included first to, potentially, flow through the whole sales process. Here, we define how a customer can get a lookup value assigned.

Sales Document

Like any other field, a **Sales Document** can inherit the `Lookup Value Code` field from a customer. It should be possible to assign a lookup value to a sales document manually. In this sub feature, we define how a sales document can get a lookup value assigned, not how it can be inherited from a customer. For that, see the **Inheritance** area.

Customer Template

In Business Central, customers can be created using so-called **Customer Templates**. This area defines how a customer template can get a lookup value assigned, not how it can be inherited to a customer. For that, see the **Inheritance** area.

Warehouse Shipment

Like on a sales document, a user should be able to assign a lookup value to a **Warehouse Shipment**. This area defines how a warehouse shipment can get a lookup value assigned, not how it can be inherited from a sales order. For that, see the **Inheritance** area.

Sales Archive

The Business Central **Sales Archive** feature allows you to archive a version of sales document. It creates a copy of the document and stores it in the `Sales Header Archive` and `Sales Line Archive` tables. In the Sales Archive sub-feature, we define how an archived sales document inherits a lookup value from its original sales document.

Inheritance

This area defines how a lookup value is inherited from a `customer` to a `Sales Document`, from a `Customer Template` to a `Customer` (including when creating a `Customer` from a `Contact`) and from a `Sales Document` to a `Warehouse Shipment`.

Posting

This area has a great resemblance to the *Inheritance* area where data flows from one entity to the other. As **Posting** is a very specific process in Business Central, it makes sense to appoint a specific area to it. This area defines how a lookup value is inherited from *Sales Document* or *Warehouse Shipment* to its posted counterpart.

Report

This area defines how a lookup value ends up on relevant reports.

Permissions

This area defines what permissions are needed to be set by the extension.

Copy Document

The Business Central **Copy Document** feature allows you create a new sales document by copying its content from any posted or non-posted sales document. In the Copy Document sub feature, we define how the new target sales document inherits a lookup value from the original source sales document.

Working all the above out in an Excel sheet, our first version of the test plan would look like *Figure 6.2* shows:

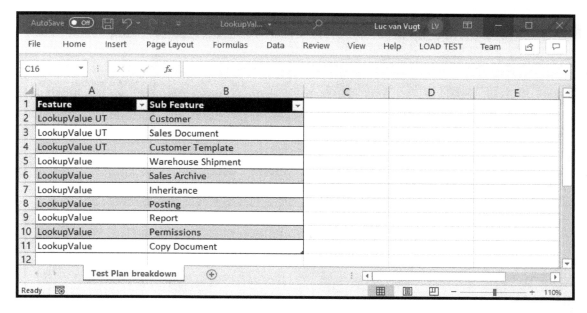

Figure 6.2 – Lookup Value test plan, first breakdown

> **Note**
>
> As you can see in *Figure 6.2*, the first three lines contain the term **UT**, for unit test. This is done as these tests are verifying whether a basic (technical) *unit* is working right and not so much a functional flow.

We could, however, go one step further, as in the context of the sub features Sales Document, Sales Archive, and Posting, for which we need to discriminate between the different (types of) documents. Therefore, a third sub feature, *Document*, is added to the test plan breakdown, even though the term document does not fully apply to all sub features.

Figure 6.3 shows the completed breakdown of our test plan from feature to scenario level. Pay attention to the consistent phrasing of each scenario, starting with a verb and describing the scenario as minimalistic as possible:

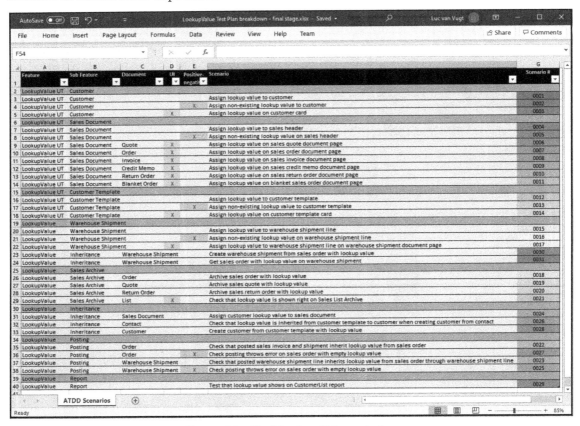

Figure 6.3 – Final Lookup Value test plan

> **Note**
>
> To be able to better read all details, please refer to the original *first* – shown in *Figure 6.2* – and *final stage* – in *Figure 6.3* – Excel sheets on the GitHub repo: `https://github.com/PacktPublishing/Automated-Testing-in-Microsoft-Dynamics-365-Business-Central-Second-Edition/tree/main/Excel sheets/Examples/Chapter 06`.

Detailing the test plan to get the test design

Given your test plan with the breakdown, we are ready to transform the test plan into the test design, detailing out each scenario using the ATDD `GIVEN`-`WHEN`-`THEN` tags. This way, we are enabling any team member to understand the different ways the software feature that needs to be implemented should behave. As we will explore various scenarios in detail with each of the test examples in this and the next chapters, we're not going to take all steps here, but give you a quick first impression. The full test design, a.k.a. ATDD sheet, is to be found on the GitHub repo here: `https://github.com/PacktPublishing/Automated-Testing-in-Microsoft-Dynamics-365-Business-Central-Second-Edition/blob/main/Excel sheets/ATDD Scenarios/LookupValue.xlsx`.

Test example 1 – a first headless test

Now, with our backpack full of tools handed over in the previous chapters, the customer wish defined, and the test plan, we're set to start creating our first automated test. Yep, it's true that the test design is also available, and that's how it should be before you start coding, but as mentioned, we will use each test example to do the test design detailing specifically.

Customer wish

Let's pick up the fundamental part of our complete customer wish: the extension of the `Customer` table with a `Lookup Value Code` field.

FEATURE

The feature we are building with our extension is called **LookupValue UT** and the specific part we are now working on is the `Customer` table. This leads to the following **FEATURE** tag:

```
[FEATURE] LookupValue UT Customer
```

SCENARIO

The specific scenario to be implemented and tested is the assignment of a lookup value to a customer, so the **SCENARIO** tag would be as follows:

```
[SCENARIO #0001] Assign lookup value to customer
```

GIVEN

The fixtures we need in order to assign a lookup value are a lookup value record and a customer record. Thus, we need the following two **GIVEN** tags:

```
[GIVEN] Lookup value
[GIVEN] Customer
```

WHEN

Given the fixture, we can set the lookup value code on the `Customer` record and therefore define our **WHEN** tag as follows:

```
[WHEN] Set lookup value on customer
```

THEN

Now that the action under test has been exercised, it is time to verify the result. Did the lookup value code field indeed get the lookup value from our fixture assigned to the customer record? This leads to the following **THEN** tag:

```
[THEN] Customer has lookup value code field populated
```

Complete scenario

So, the complete scenario definition would then be to allow us to copy it later when creating our test code:

```
[FEATURE] LookupValue Customer
[SCENARIO #0001] Assign lookup value to customer
[GIVEN] Lookup value
[GIVEN] Customer
[WHEN] Set lookup value to customer
[THEN] Customer has lookup value code field populated
```

> **Note**
> The preceding lines will literally be pasted into a test function as comments, where they will act as a guide to write the actual test code.

Application code

The first part of the customer wish, that is, `[SCENARIO #0001]`, defines the need for a `LookupValue`, which is a new table from which a value can be assigned to a customer by means of a so-called `Lookup Value Code` field. This has been implemented by means of the following `.al` objects:

```
table 50000 "LookupValue"
{
  LookupPageId = "LookupValues";
```

```
    fields
    {
      field(1; Code; Code[10]){}
      field(2; Description; Text[50]){}
    }

    keys
    {
      key(PK; Code)
      {
        Clustered = true;
      }
    }
}
page 50000 "LookupValues"
{
    PageType = List;
    SourceTable = "LookupValue";

    layout
    {
      area(content)
      {
        repeater(RepeaterControl)
        {
          field("Code"; "Code"){}
          field("Description"; "Description"){}
        }
      }
    }
}

tableextension 50000 "CustomerTableExt" extends Customer
{
    fields
    {
      field(50000; "Lookup Value Code"; Code[10])
```

```
    {
      TableRelation = "LookupValue";
    }
  }
}


pageextension 50000 "CustomerCardPageExt" extends "Customer Card"
{
  layout
  {
    addlast(General)
    {
      field("Lookup Value Code"; "Lookup Value Code"){}
    }
  }
}
```

In the application code, the bare minimum has been included to save space. Properties like `Caption`, `ApplicationArea`, `DataClassification`, `UsageCategory`, and `ToolTip` have been left out. Download the **LookupValue** extension from GitHub to get the complete objects.

Test code

With the first part of the customer wish clear, we have a neat structure to start writing our first test.

Steps to take

The following are the steps to take:

1. Create a test codeunit with a name based on the **FEATURE** tag.

2. Embed the customer wish into a test function with a name based on **SCENARIO** tag.

3. Write your test story, based upon **GIVEN**, **WHEN**, and **THEN** tags.

4. Construct your real code.

This procedure is what we will be utilizing when implementing the test code and what we will call from now on the **4-steps recipe**.

Create a test codeunit

Using the **FEATURE** tag's name, the basic structure of our codeunit will be like this:

```
codeunit 81000 "LookupValue UT Customer"
{
  Subtype = Test;

  trigger OnRun()
  begin
    //[FEATURE] LookupValue UT Customer
  end;
}
```

As you can see, the UT added to the **FEATURE** tag stands for Unit Test, marking that the tests are unit tests and not functional tests. As an easy start, `codeunit 81000` already exists in the **LookupValue** extension on GitHub.

Embed the customer wish into a test function

Now, we create a test function with a name based on the **SCENARIO** description and embed the customer wish, **GIVEN**-**WHEN**-**THEN**, in this function.

I call this "embedding the *green*," being the **GIVEN**-**WHEN**-**THEN** comment sentences, before you start programming the *black*, being the .al test code. Look at what the codeunit has now become:

```
codeunit 81000 "LookupValue UT Customer"
{
  Subtype = Test;

  trigger OnRun()
  begin
    //[FEATURE] LookupValue UT Customer
  end;

  [Test]
  procedure AssignLookupValueToCustomer()
  begin
    //[SCENARIO #0001] Assign lookup value to customer
    //[GIVEN] Lookup value
```

```
//[GIVEN] Customer
//[WHEN] Set lookup value on customer
//[THEN] Customer has lookup value code field populated
end;
}
```

> **Note**
>
> With the possibility of selecting a color theme in VS Code, "embedding the *green*" and "programming the *black*" might not be as obvious to all readers as comment lines might be a different color in VS Code than the green I am referring to, and code lines might not appear black. I, however, still love to mention it like this, especially when I am teaching or presenting and showing my code in an almost old-fashioned way using the **Light** (**Visual Studio**) color theme. Also note that the code style in this book does not allow for green comment lines.

Write your test story

For me, writing the first *black* parts is about writing pseudo-English, defining **what** I need to achieve with my test. It makes my test readable by any non-technical person in the project, and, if I need their support, the threshold for them to read the test is substantially lower than when I would have immediately started to write the *real* thing. And maybe an even stronger argument – I will have my code embedded in reusable helper functions.

So here we go, let's write the *black* parts:

```
codeunit 81000 "LookupValue UT Customer"
{
    Subtype = Test;

    trigger OnRun()
    begin
        //[FEATURE] LookupValue UT Customer
    end;

    [Test]
    procedure AssignLookupValueToCustomer()
    begin
        //[SCENARIO #0001] Assign lookup value to customer

        //[GIVEN] Lookup value
```

```
      CreateLookupValueCode();
      //[GIVEN] Customer
      CreateCustomer();

      //[WHEN] Set lookup value on customer
      SetLookupValueOnCustomer();

      //[THEN] Customer has lookup value code field populated
      VerifyLookupValueOnCustomer();
    end;
  }
```

With this story, the **what**, I have been designing four helper functions with no argument(s) and return value yet. These helper functions implement the **how** of my test and will be defined in the next step. This way, we decouple our domain logic, the logical test, from the implementation details. This is real programming code but written as a function call that has an easy-to-read, non-technical name. Written like this, they look like easy steps in a process that any technical or non-technical person should be able to interpret. All members of the test team should be able to read through the test functions and understand the steps. The "technical" programming code will be inside those functions, as you will see in the next steps.

Note how close the names of the helper functions are to the description of the tag it belongs to.

Note

Without being aware, my *writing your test story* pattern is a known refactoring pattern: the **compose method** pattern. One of my interviewees pointed this out.

Here you can find a nice description of the **compose method** refactoring pattern: https://scrutinizer-ci.com/docs/refactorings/compose-method.

Construct the real code

If you're a developer, I might have been challenging you with my pseudo code up until this point, using no real code but only a structure. Now, ready yourself as the real part starts right now, and I hope, for you and your teammates, you will do the same with the tests you will be coding yourself in the future.

While inspecting our first test function, I already concluded that I need the following four helper functions:

- CreateLookupValueCode
- CreateCustomer
- SetLookupValueOnCustomer
- VerifyLookupValueOnCustomer

Let's construct and discuss these.

CreateLookupValueCode

CreateLookupValueCode is a reusable helper function to create a pseudo-random LookupValue record. In a later stage, we could promote this to a to-be-created library codeunit:

```
local procedure CreateLookupValueCode(): Code[10]
var
  LookupValue: Record LookupValue;
begin
  LookupValue.Init();
  LookupValue.Validate(
    Code,
    LibraryUtility.GenerateRandomCode(
        LookupValue.FieldNo(Code),
        Database::LookupValue));
  LookupValue.Validate(Description, LookupValue.Code);
  LookupValue.Insert();
  exit(LookupValue.Code);
end;
```

To populate the PK field, we make use of the GenerateRandomCode function from the standard test library, Library - Utility, codeunit 131000. The LibraryUtility variable will be declared globally as Microsoft does in their test codeunits, making it reusable in other helper functions.

> **Notes**
>
> (1) Pseudo-random means that, whenever our test is executed in the same context, the GenerateRandomCode function will yield the same value, contributing to a reproducible test.
>
> (2) The Description field is populated by the same value as the Code field as the specific value of Description is of no meaning, and this way it is the most effective.

CreateCustomer

Using the CreateCustomer function, from the standard library codeunit Library – Sales (130509), our CreateCustomer creates a useable customer record and makes this helper function a straightforward exercise:

```
local procedure CreateCustomer(var Customer: record
  Customer)
begin
  LibrarySales.CreateCustomer(Customer);
end;
```

As with the preceding LibraryUtility variable, we will declare the LibrarySales variable globally, so we can use the same variable throughout this test codeunit.

You might wonder why we create a helper function that only has one statement line. As mentioned, using helper functions makes the test readable for non-technical people as well as making it reusable. However, it also makes it more maintainable and easier to extend by adding one or more lines of code. If we need to add an update to the customer record created by the CreateCustomer function in the Library - Sales codeunit, we only need to add that to our local CreateCustomer function.

> **Note**
>
> As a general rule, I try not to call library functions directly from test functions. This comes with some exceptions, as we will see later.

SetLookupValueOnCustomer

Have a look at the implementation of SetLookupValueOnCustomer:

```
local procedure SetLookupValueOnCustomer(
  var Customer: record Customer; LookupValueCode: Code[10])
begin
  Customer.Validate("Lookup Value Code", LookupValueCode);
```

```
    Customer.Modify();
  end;
```

Calling `Validate` is essential here. `SetLookupValueOnCustomer` is not just about assigning a value to the `Lookup Value Code` field, but also about making sure it is validated against the existing values in the `LookupValue` table. Note that the `OnValidate` trigger of the `Lookup Value Code` field does not contain code.

> **Note**
>
> I am often asked why I am using `Modify()` and not `Modify(true)` in the previous helper function (`SetLookupValueOnCustomer`). The sole reason is that a default `Modify()`, being actually `Modify(false)`, will not trigger the `OnModify` trigger. This saves execution time. Of course, this can only be done if the code that resides in the `OnModify` trigger does not need to be executed. As long as we can *fake it*, we will.
>
> The same applies to validating a record field: I will only validate a field when the code in the `OnValidate` trigger of a field should be triggered.

VerifyLookupValueOnCustomer

As mentioned in *Chapter 5, Test Plan and Test Design, a test without verification is no test at all,* so we need to verify that the lookup value code assigned to the `Lookup Value Code` field of the customer record is indeed the value that was created in the `Lookup Value` table. We, therefore, retrieve the record from the database as follows:

```
local procedure VerifyLookupValueOnCustomer(
  CustomerNo: Code[20]; LookupValueCode: Code[10])
var
  Customer: Record Customer;
  FieldOnTableTxt: Label '%1 on %2';
begin
  Customer.Get(CustomerNo);
  Assert.AreEqual(
    LookupValueCode,
    Customer."Lookup Value Code",
    StrSubstNo(
      FieldOnTableTxt,
      Customer.FieldCaption("Lookup Value Code"),
      Customer.TableCaption()))
```

```
        );
    end;
```

To verify that the expected value (first argument) and the actual value (second argument) are equal, we make use of the `AreEqual` function in the standard library codeunit `Library Assert`, codeunit 130002. Of course, we could build our own verification logic using the `Error` system function, but hey, that's what `AreEqual` is doing, and some more. Have a look:

```
procedure AreEqual(Expected: Variant; Actual: Variant; Msg: Text)
begin
    if not Equal(Expected, Actual) then
        Error(
            AreEqualFailedErr,
            Expected,
            TypeNameOf(Expected),
            Actual,
            TypeNameOf(Actual),
            Msg)
end;
```

By using the `AreEqual` function, we ensure that we get a standardized error message in case the expected and actual values are not equal. For an example, see *Figure 6.5*. Over time, by reading the error of any failing test where the verification helper functions are making use of `Library Assert` methods, you will be able to recognize easily what kind of error occurred. Using the third argument of the `AreEqual` function, the Text data type `Msg`, will allow you to add some extra context and thereby help you to more easily identify the cause of the failure.

> **Note**
> `Library Assert` (130002) is a replacement of the older library codeunit `Assert` (130000). It's one of the first steps of Microsoft replacing their long-existing test collateral (including the libraries) with new, refactored codeunits.

Our complete test codeunit would look like the following code that is ready for execution. Notice the variables and arguments added to the test codeunit and functions:

```
codeunit 81000 "LookupValue UT Customer"
{
    Subtype = Test;

    trigger OnRun()
```

```
begin
  //[FEATURE] LookupValue UT Customer
end;

var
  Assert: Codeunit "Library Assert";
  LibraryUtility: Codeunit "Library - Utility";
  LibrarySales: Codeunit "Library - Sales";

[Test]
procedure AssignLookupValueToCustomer()
var
  Customer: Record Customer;
  LookupValueCode: Code[10];
begin
  //[SCENARIO #0001] Assign lookup value to customer

  //[GIVEN] Lookup value
  LookupValueCode := CreateLookupValueCode();
  //[GIVEN] Customer
  CreateCustomer(Customer);

  //[WHEN] Set lookup value on customer
  SetLookupValueOnCustomer(Customer, LookupValueCode);

  //[THEN] Customer has lookup value code field populated
  VerifyLookupValueOnCustomer(Customer."No.", LookupValueCode);
end;

local procedure CreateLookupValueCode(): Code[10]
var
  LookupValue: Record LookupValue;
begin
  LookupValue.Init();
  LookupValue.Validate(
    Code,
    LibraryUtility.GenerateRandomCode(
```

```
      LookupValue.FieldNo(Code),Database::LookupValue));
    LookupValue.Validate(Description, LookupValue.Code);
    LookupValue.Insert();
    exit(LookupValue.Code);
  end;


  local procedure CreateCustomer(var Customer: record
    Customer)
  begin
    LibrarySales.CreateCustomer(Customer);
  end;


  local procedure SetLookupValueOnCustomer(
    var Customer: record Customer; LookupValueCode:
      Code[10])
  begin
    Customer.Validate("Lookup Value Code",LookupValueCode);
    Customer.Modify();
  end;


  local procedure VerifyLookupValueOnCustomer(
    CustomerNo: Code[20]; LookupValueCode: Code[10])
  var
    Customer: Record Customer;
    FieldOnTableTxt: Label '%1 on %2';
  begin
    Customer.Get(CustomerNo);
    Assert.AreEqual(
      LookupValueCode,
      Customer."Lookup Value Code",
      StrSubstNo(
        FieldOnTableTxt,
        Customer.FieldCaption("Lookup Value Code"),
        Customer.TableCaption())
      );
  end;
}
```

Test execution

Now that we are ready with our first test, we can deploy the **LookupValue** extension to our Dynamics 365 Business Central installation. If we set the test tool page as the start object in the launch.json, we can immediately add our test codeunit to the DEFAULT suite as follows:

```
"startupObjectType": "Page","startupObjectId": 130451
```

Running the test by selecting the **Run** action will show that it executes successfully: green.

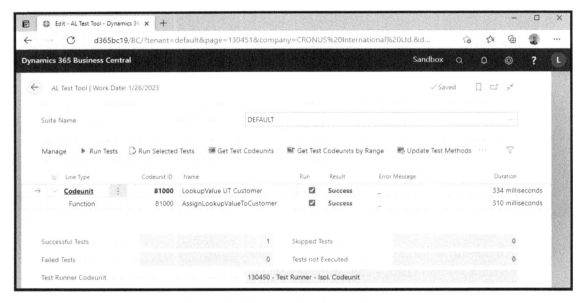

Figure 6.4 – Test example 1 having been run

Congratulations, we have implemented our first successful test, as *Figure 6.4* shows!

Test the test

During my workshop, attendees challenge me and ask: "Indeed the test result is successful, but how do I know success is a real success? How can I test the test?"

The following two options are proven ways to *test the test*:

- Test the data being created.
- Adjust the test so the verification errs.

Test the data being created

Testing the data being created can be done in two ways:

- Run the test outside of a test isolation and have a look at the Customer table. Find out that a new customer has been created with the Lookup Value Code field populated, and, of course, a related record in the Lookup Value table.

- Debug your test and, using SQL Server Management Studio, run SQL queries on the Customer and Lookup Value table. Make sure that you read uncommitted data to find the same records before the test is finished.

The latter is my preferred method as it makes it possible to run tests in isolation, thus not changing the database irreversibly. It also allows us to see what data is being created.

> **Note**
>
> The second option, *Debug your test and run SQL queries*, is only possible in an on-premises or containerized installation of Dynamics 365 Business Central because you cannot directly access the Azure SQL database for any SaaS tenant.
>
> I will make use of this method in *Chapter 11, How to Construct Complex Scenarios*. However, if you already want to learn more about this, please refer to my Areopa webinar *How about verifying and debugging your tests* on YouTube: https://www.youtube.com/watch?v=zna9iKllz-o.

Adjust the test so the verification errs

This is maybe the easiest and most robust option: make sure the verification fails by providing another value for the expected result. In the case of our test, for example, provide your own name:

```
//[THEN] Customer has lookup value code field populated
VerifyLookupValueOnCustomer(Customer."No.", 'LUC');
```

Running the test again shows the test indeed fails on the verification part, as the **Call Stack** clearly shows in *Figure 6.5*.

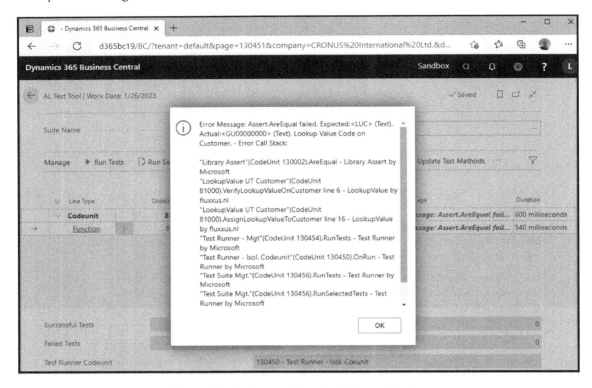

Figure 6.5 – Test example 1 with failing verification

The error thrown tells us that the expected value is LUC while the actual value is GU00000000:

```
Assert.AreEqual failed. Expected:<LUC> (Text). Actual:<GU00000000>
(Text). Lookup Value Code on Customer.
```

As mentioned before, **Error Message** also displays the Call Stack of the execution that leads to this error.

> **Note**
> Having your code under source code management and committed to the first version of your tests, it's fairly easy to *test the test*. Once you have let the verification err, and studied and approved its result, just discard this temporary change.

Test example 2 – a first positive-negative test

This test example does not relate to a new customer wish and no new application code but is complementary to our first test. It is a *rainy* path version of the same customer wish leading to a new scenario:

```
[FEATURE] LookupValue UT Customer
[SCENARIO #0002] Assign non-existing lookup value to customer
[GIVEN] Non-existing lookup value
[GIVEN] Customer
[WHEN] Set non-existing lookup value on customer
[THEN] Non existing lookup value error thrown
```

Test code steps

Let's reuse the recipe applied for test example 1.

You might already recall these are the steps to take, being our new tool, the *4-steps recipe*:

1. Create a test codeunit, with a name based on the **FEATURE** tag.
2. Embed the customer wish into a test function with a name based on the **SCENARIO** tag.
3. Write your test story, based on **GIVEN**, **WHEN**, and **THEN** tags.
4. Construct your real code.

Create a test codeunit

Sharing the same **FEATURE** tag values as test example 1, our new test case will share the same test codeunit, that is, `codeunit 81000 LookupValue UT Customer`.

Embed the customer wish into a test function

Embedding the green results in the following new test function in `codeunit 81000`:

```
[Test]
procedure AssignNonExistingLookupValueToCustomer()
begin
  //[SCENARIO #0002] Assign non-existing lookup value to
  //                 customer
  //[GIVEN] Non-existing lookup value
  //[GIVEN] Customer
  //[WHEN] Set non-existing lookup value on customer
```

```
    //[THEN] Non existing lookup value error thrown
end;
```

Write your test story

Filling in the first *black* **story elements** leads to the following typical choices:

- Creating a non-existing lookup value is achieved by just providing a string constant that has no related record in the Lookup Value table.

- To assign this value to the Lookup Value Code field on a customer, we do not need a customer record in the database. A local variable suffices to trigger the error we want to happen.

- Setting the lookup value on the customer can be achieved by using the SetLookupValueOnCustomer from test example 1. Assigning a non-existing value will result in an error. The objective of the test is to verify that an error will happen, so we add the asserterror keyword before the function call to make this a successful test.

As a consequence, the test story already has some more details than our previous test:

```
procedure AssignNonExistingLookupValueToCustomer()
var
    Customer: Record Customer;
    LookupValueCode: Code[10];
begin
    //[SCENARIO #0002] Assign non-existing lookup value to
    //                 customer

    //[GIVEN] Non-existing lookup value
    LookupValueCode := 'SC #0002';
    //[GIVEN] Customer record variable
    // See local variable Customer

    //[WHEN] Set non-existing lookup value on customer
    asserterror SetLookupValueOnCustomer(Customer, LookupValueCode);

    //[THEN] Non existing lookup value error thrown
    VerifyNonExistingLookupValueError(LookupValueCode);
end;
```

> **Note**
>
> Observe how we *keep it simple and yet thorough*, as discussed in the previous chapter.

Construct the real code

We can re-use the SetLookupValueOnCustomer function that we created in the previous test, and we only need to create one new helper function: VerifyNonExistingLookupValueError.

VerifyNonExistingLookupValueError

Like in our first verification function, we make use of a function from the standard library codeunit Library Assert (130002) called ExpectedError. We only need to provide ExpectedError the expected error text. The actual error will be retrieved by ExpectedError using the GetLastErrorText system function:

```
local procedure VerifyNonExistingLookupValueError(
  LookupValueCode: Code[10])
var
  Customer: Record Customer;
  LookupValue: Record LookupValue;
  ValueCannotBeFoundInTableTxt: Label
    'The field %1 of table %2 contains a value (%3) that cannot
      be found in the related table (%4).';
begin
  Assert.ExpectedError(
    StrSubstNo(
      ValueCannotBeFoundInTableTxt,
      Customer.FieldCaption("Lookup Value Code"),
      Customer.TableCaption(),
      LookupValueCode,
      LookupValue.TableCaption()));
end;
```

Note how the expected error text is constructed by using the StrSubstNo system method in conjunction with the ValueCannotBeFoundInTableTxt label.

Test execution

Let's redeploy our extension and add the second test to the test tool by selecting **Update Test Methods**. **Update Test Methods** will update the selected test codeunit by adding all current test functions residing in the codeunit as lines to the test tool. Note that the **Result** column will be cleared. Now, run the test codeunit and see, as shown in *Figure 6.6*, that both tests are successful and colored `green`:

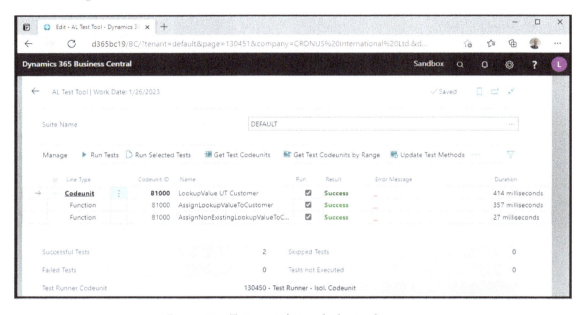

Figure 6.6 – Test example 1 and 2 having been run

Test the test

How to verify that the success is a real success? You can do this in a similar way as you did with test example 1: provide a different expected value to the verification function of our test case. So, let's do it.

Adjust the test so the verification errs

Adjust the verification in a similar way as you did in test example 1:

```
// [THEN] Non existing lookup value error thrown
VerifyNonExistingLookupValueError('LUC');
```

Deploy the extension again and run the second test only by selecting the test function line and executing **Run Selected Tests**. The output is shown in *Figure 6.7*.

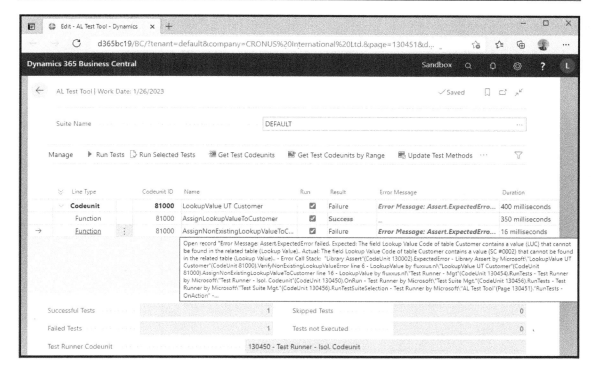

Figure 6.7 – Failing verification of test example 2

The error thrown tells us that the expected value is LUC while the actual value is SC #0002 as follows:

```
Assert.ExpectedError failed.
 Expected: The field Lookup Value Code of table Customer contains
a value (LUC) that cannot be found in the related table (Lookup
Value)..
 Actual: The field Lookup Value Code of table Customer contains a
value (SC #0002) that cannot be found in the related table (Lookup
Value)..
```

> **Note**
>
> As you can see in *Figure 6.7*, instead of clicking on the hyperlink of the error message, you can also hover over the error message to see the full error and the related Call Stack.

Removing asserterror

With *rainy* path scenarios, we typically use an `asserterror` to wrap the `WHEN` part to catch the error; consequently, we have another way to test the test: by removing `asserterror` and running the test again. Now, you will get to see the error triggered by the `WHEN` part:

```
The field Lookup Value Code of table Customer contains a value (SC
#0002) that cannot be found in the related table (Lookup Value).
```

This is the `Non existing lookup value error thrown` message we verify against in the `THEN` part. Triggering this error message and copy/pasting it is a simple way to build in `.al`, the expected error label `ValueCannotBeFoundInTableTxt`.

> **Notes**
>
> (1) The `Modify` in `SetLookupValueOnCustomer` strictly is not possible in test example 2 as we do not retrieve a customer record from the database. The error thrown by `Validate`, however, will prevent `Modify` from being called.
>
> (2) Test example 2 is built upon the assumption that the `TableRelation` set on the `Lookup Value Code` field on the `Customer` table uses the default setting of the `ValidateTableRelation` property on that field.

Test example 3 – a first UI test

The field `Lookup Value Code` has been implemented on the `Customer` table and tested. But, of course, to allow users to manage it, it needs to be made accessible in the UI. Consequently, it needs to be placed on the `Customer Card`.

Customer wish

The next stage of the customer wish is very close to the previous part defined by [`SCENARIO #0001`]. The main difference is that we now want to access a customer by means of the **UI element** `Customer Card`. By mimicking end users, our scenario describes creating a new `Customer Card` (see second `GIVEN`). Have a look:

```
[FEATURE] LookupValue UT Customer
[SCENARIO #0003] Assign lookup value on customer card
[GIVEN] Lookup value
[GIVEN] Customer card
[WHEN] Set lookup value on customer card
[THEN] Customer has lookup value code field populated
```

Test code

Consecutively, let's step through our **4-steps recipe**.

Create a test codeunit

Again, sharing the same **FEATURE** tag values as in our previous tests, we can place our new test case in the same test codeunit, that is, codeunit `LookupValue UT Customer (81000)`.

Embed the customer wish into a test function

Wrap the *green* into a new test function in codeunit 81000 as follows:

```
[Test]
procedure AssignLookupValueToCustomerCard()
begin
    //[SCENARIO #0003] Assign lookup value on customer card
    //[GIVEN] Lookup value
    //[GIVEN] Customer card
    //[WHEN] Set lookup value on customer card
    //[THEN] Customer has lookup value field populated
end;
```

Write your test story

A straightforward version of the new test story, in parallel with test example 1, would be:

```
[Test]
procedure AssignLookupValueToCustomerCard()
begin
    //[SCENARIO #0003] Assign lookup value on customer card

    //[GIVEN] Lookup value
    CreateLookupValueCode();
    //[GIVEN] Customer card
    CreateCustomerCard();

    //[WHEN] Set lookup value on customer card
    SetLookupValueOnCustomerCard();
```

```
    //[THEN] Customer has lookup value field populated
    VerifyLookupValueOnCustomer();
  end;
```

Adding variables and arguments, the code becomes:

```
  [Test]
  procedure AssignLookupValueToCustomerCard()
  var
    CustomerCard: TestPage "Customer Card";
    CustomerNo: Code[20];
    LookupValueCode: Code[10];
  begin
    //[SCENARIO #0003] Assign lookup value on customer card

    //[GIVEN] Lookup value
    LookupValueCode := CreateLookupValueCode();
    //[GIVEN] Customer card
    CreateCustomerCard(CustomerCard);

    //[WHEN] Set lookup value on customer card
    CustomerNo := SetLookupValueOnCustomerCard(
      CustomerCard, LookupValueCode);

    //[THEN] Customer has lookup value field populated
    VerifyLookupValueOnCustomer(CustomerNo, LookupValueCode);
  end;
```

To access the UI in automated tests, we need to make use of the 5th pillar: the TestPage. As you can see in the specific case of our test function AssignLookupValueToCustomerCard, the test page object is based on the Customer Card page.

Construct the real code

We can make use of the already existing helper functions `CreateLookupValueCode` and `VerifyLookupValueOnCustomer`, but we also need to construct the following two new helper functions:

- `CreateCustomerCard`
- `SetLookupValueOnCustomerCard`

CreateCustomerCard

To create a new customer card, we just call the `OpenNew` method that any editable `TestPage` has to its availability:

```
local procedure CreateCustomerCard(
  var CustomerCard: TestPage "Customer Card")
begin
  CustomerCard.OpenNew();
end;
```

SetLookupValueOnCustomerCard

Using the control method `SetValue` we set the value of the `Lookup Value Code` field and return the number of the created customer – `CustomerNo` – needed for the verification later:

```
local procedure SetLookupValueOnCustomerCard(
  var CustomerCard: TestPage "Customer Card";
  LookupValueCode: Code[10]) CustomerNo: Code[20]
begin
    CustomerCard."Lookup Value Code".SetValue(LookupValueCode);
    CustomerNo := CustomerCard."No.".Value();
    CustomerCard.Close();
end;
```

As `SetValue` mimics the user setting a value, it will trigger the validation of the field. If a non-existing value is entered, it will validate the value against the existing records in the `Lookup Value` table as we tested in *test example 2*. To retrieve the value of the `No.` field, you use the control method `Value` and need to close the page to trigger the system, so save the changes to the record to the database.

> **Note**
>
> The `Value` method has a twofold purpose. It can be used to get or set the value of a field (control). The difference between setting the value using `Value` or `SetValue` is that `Value` always takes a string as an argument, while the argument of `SetValue` should be the same data type as the data type of the field (control).

You're almost ready to run the new test. There is, however, one major thing failing in the `SetLookupValueOnCustomerCard` helper function. It will operate fine, but it does not take account for, in my opinion, a design flaw: `SetLookupValueOnCustomerCard` will run successfully, even when the `Lookup Value Code` field is not editable. Both `SetValue` and `Value` do not check on this. As the whole purpose of our test is to check whether the user can set the `Lookup Value Code` field, we need to add a small verification to determine whether the field is editable. Because of this, the `SetLookupValueOnCustomerCard` function needs to be updated to the following using another function from the `Library Assert` codeunit `IsTrue`:

```
local procedure SetLookupValueOnCustomerCard(
  var CustomerCard: TestPage "Customer Card";
    LookupValueCode: Code[10]) CustomerNo: Code[20]
begin
    Assert.IsTrue(
      CustomerCard."Lookup Value Code".Editable(),
        'Editable');
    CustomerCard."Lookup Value Code".SetValue(LookupValueCode);
    CustomerNo := CustomerCard."No.".Value();
    CustomerCard.Close();
end;
```

> **Note**
>
> Applying both `Value` and `SetValue` to a non-visible field will throw an error.

Test execution

Let's again redeploy the extension and add the new test function using the **Update Test Methods** feature, and run the tests: **red. Red**? Yes, **red**; see *Figure 6.8*.

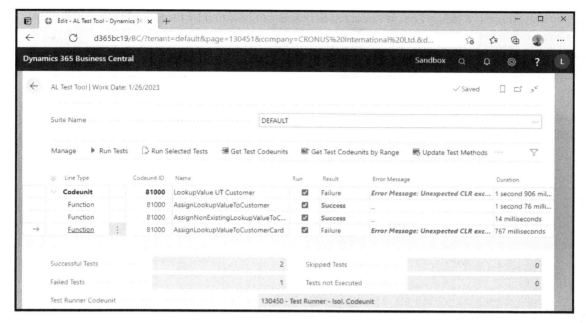

Figure 6.8 – Failing test example 3

Shoot, an error has occurred. Error? Let's click on the hyperlink, read the error, and try to understand:

```
Unexpected CLR exception thrown.: Microsoft.Dynamics.
Framework.UI. FormAbortException: Page New - Customer Card
has to close ---> Microsoft. Dynamics.Nav.Types.Exceptions.
NavNCLMissingUIHandlerException: Unhandled UI: ModalPage 1380 --->
System.Reflect
```

I must confess, I always get a bit nervous of those techie **CLR exception thrown** messages, but I have learned to scan for the things that relate to what I know. Here are two things:

- `NavNCLMissingUIHandlerException`
- `Unhandled UI: ModalPage 1380`

Apparently, there is a `ModalPage` occurrence that is not handled by our test. More specifically it is page 1380 that's called modally and not handled. Page 1380? It's the `Config Templates` page, the one that pops up when you are going to create a new customer, as shown in *Figure 6.9*.

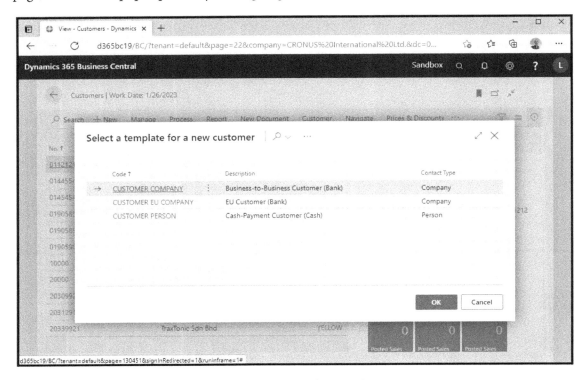

Figure 6.9 – Page 1380 popping up when creating a new customer

So, we need to construct a `ModalPageHandler` and link that to our third test:

```
[ModalPageHandler]
procedure HandleCustomerTemplList(
  var CustomerTemplList: TestPage "Select Customer Templ. List")
begin
    CustomerTemplList.OK.Invoke();
end;
```

Set the `HandlerFunctions` tag to the test function with reference to the `ModalPageHandler`:

```
[Test]
[HandlerFunctions('HandleCustomerTemplList')]
procedure AssignLookupValueToCustomerCard()
```

Now the test runs successfully: **green**!

> **Note**
>
> As *Appendix, Getting Up and Running with Business Central, VS Code, and the GitHub Project*, discusses, the GitHub repo has two branches, **main** and **BC18**, where **main** contains the code based upon BC19 and **BC18**, indeed, the code based on BC18. The reason is that the customer template feature has been somewhat redesigned and as a consequence, the customer template list triggered by our test is another one. On BC19 it is page 1380, `Select Customer Templ. List`, and in BC18 – and before – it is page 1340, `Config Templates`.

Test the test

Let's *test the test* and verify that it is a good test.

Adjust the test so the verification errs

A proven method can be achieved in exactly the same way as we have done for test example 1. Next to that note, we have added another verification to our code:

```
Assert.IsTrue("Lookup Value Code".Editable(), 'Editable');
```

Change `IsTrue` into `IsFalse`. You will see that the test fails as the `Lookup Value Code` field is editable. The `IsTrue` verification ensures that, when the `Lookup Value Code` field is turned to non-editable, the test will fail.

Congrats, you have built your first set of automated tests. Before we continue with a next set in the following chapter, let's have a short discussion on headless and UI testing.

Headless versus UI

As mentioned before, **headless** testing is the preferred mode for automated tests as it is faster than **UI testing**. With *test example 1* and *3*, we did implement the same kind of test: check that a lookup value can be assigned to a customer. *Test example 1* is in headless mode, while *test example 3* uses the UI. Running both tests indeed shows that UI tests are slower than headless tests, measured over 12 different runs. Have a look at the graph of execution duration (in seconds) in *Figure 6.10*:

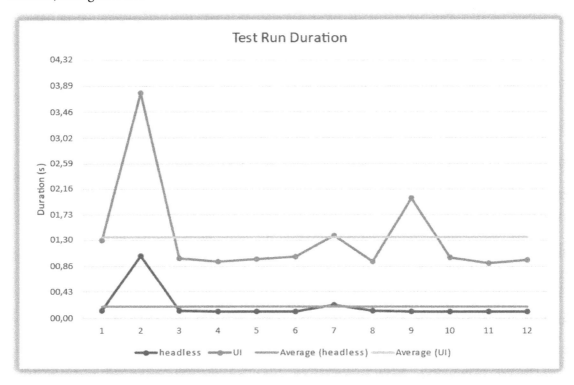

Figure 6.10 – Test run duration of headless vs. UI tests

The average execution duration for the UI tests is 1.35 seconds, while the headless average is almost 7 times faster: 0.20 seconds.

> **Note**
> Recently an attendee of one of my online courses showed me an example of two of his tests achieving the same: one in headless mode and the other using the UI. In this specific case, the latter was even 50 times slower!

Summary

In this chapter, you got to build your first automated tests and create a test plan for your customer wish and detailed it out into a test design utilizing the ATDD test case pattern to design each test. With this ATDD *base structure* and the *4-steps recipe*, you learned how to create a test codeunit, embed the customer wishes into a test, write a test story, and finally construct your real code. Next to these principles, we paid attention to writing positive-negative and UI tests.

In the next chapter, *Chapter 7, From Customer Wish to Test Automation – Next Level*, we will continue to use ATDD and the 4-steps recipe to create some more advanced tests.

7

From Customer Wish to Test Automation – Next Level

In the previous chapter, you built your first basic test automation in Dynamics 365 Business Central. You looked at three simple examples that show how to apply the **Acceptance Test-Driven Development (ATDD)** test case pattern and our *4-steps recipe* to get customer wishes converted into an application and test code. In this chapter, you will use the same methodology to create some more tests that use the same test data setup, that have a comparable structure, and that interact with UI handlers.

The following topics will be covered:

- Sales documents, customer template, and warehouse shipment
- Test example 4 – how to set up a shared fixture
- Test example 5 – how to parameterize tests
- Test example 6 – how to hand over data to UI handlers

Technical requirements

All test example code can be found on GitHub: `https://github.com/PacktPublishing/Automated-Testing-in-Microsoft-Dynamics-365-Business-Central-Second-Edition`. In this chapter, we will continue with the code from *Chapter 6*, *From Customer Wish to Test Automation – the Basics*: `https://github.com/PacktPublishing/Automated-Testing-in-Microsoft-Dynamics-365-Business-Central-Second-Edition/tree/main/Chapter 06 (LookupValue Extension)`.

The final code for this chapter can be found in `https://github.com/PacktPublishing/Automated-Testing-in-Microsoft-Dynamics-365-Business-Central-Second-Edition/tree/main/Chapter 07 (LookupValue Extension)`.

Details on how to use this repository and how to set up VS Code are discussed in *Appendix, Getting Up and Running with Business Central, VS Code, and the GitHub Project*.

Sales documents, customer template, and warehouse shipment

With the three examples in *Chapter 6*, *From Customer Wish to Test Automation – the Basics*, we added the `Lookup Value Code` field to the `Customer` table. However, that's just a part of the customer wish as it clearly describes that:

"… this field has to be carried over to the whole bunch of sales documents and, at the same time, it needs to be included in the warehouse shipping."

So, before we dive into the following test examples, a note needs to be made that parallel to the implementation of the `Lookup Value Code` field on the `Customer` table, the same field has to be implemented on the `Sales Header` table, the `Customer Template` table, the `Warehouse Shipment Line` table, and all related pages. The ATDD test case descriptions are very much alike, and this will be the same for the application and test code. Copy and paste – the great virtue of any Business Central developer.

Let's have a look at what the ATDD test case descriptions look like for the customer template:

```
[SCENARIO #0012] Assign lookup value to customer template
[GIVEN] Lookup value
[GIVEN] Customer template
[WHEN] Set lookup value on customer template
[THEN] Customer template has lookup value code field populated
[SCENARIO #0013] Assign non-existing lookup value to customer
```

```
                    template
[GIVEN] Non-existing lookup value
[GIVEN] Customer template record variable
[WHEN] Set non-existing lookup value to customer template
[THEN] Non existing lookup value error was thrown
[SCENARIO #0014] Assign lookup value on customer template card
[GIVEN] Lookup value
[GIVEN] Customer template card
[WHEN] Set lookup value on customer template card
[THEN] Customer template has lookup value code field populated
```

Do you see the resemblance with scenarios #0001, #0002, and #0003, that we built in *Chapter 6, From Customer Wish to Test Automation – the Basics?* On GitHub, you will find both the full list of ATDD scenarios and the complete test code.

Test example 4 – how to set up a shared fixture

Although it isn't explicitly mentioned, we created a **fresh fixture** for each of the three previous tests as defined as per the **GIVEN** tags, a lookup-value record, and a customer record. For speed purposes, however, it does make sense to consider whether you need a fresh fixture for each test or a **shared fixture** for a group of tests. In the case of scenarios #0001 and #0003, we could perfectly do with the same LookupValueCode, no need to create a new lookup value record for each of these tests.

Customer wish

Let's use the part of the customer wish that prescribes to have a Lookup Value Code field on all sales documents to illustrate how a shared fixture can be achieved. This would come down to the following eight scenarios, leaving out the **GIVEN**-**WHEN**-**THEN** part to save space:

```
[SCENARIO #0004] Assign lookup value to sales header
[SCENARIO #0005] Assign non-existing lookup value on sales header
[SCENARIO #0006] Assign lookup value on sales quote document page
[SCENARIO #0007] Assign lookup value on sales order document page
[SCENARIO #0008] Assign lookup value on sales invoice document
                 page
[SCENARIO #0009] Assign lookup value on sales credit memo
                 document page
```

```
[SCENARIO #0010] Assign lookup value on sales return order
                 document page
[SCENARIO #0011] Assign lookup value on blanket sales order
                 document page
```

With *Chapter 6*, *From Customer Wish to Test Automation – the Basics*, fresh in your mind, you might notice that scenarios #0001, #0004, and #0012 are quite similar. This is the same for scenarios #0003 and #0006 through #0011 and #0014. All of these scenarios share the following same **GIVEN** part:

```
[GIVEN] Lookup value
```

Straightforward implementation of this requirement would lead to creating a lookup value record seven times. So, we'll be *lazy* and apply the shared fixture, or lazy setup, pattern, and of course, at the same time making it easier to maintain.

Application code

This part of the customer wish leads to the implementation of the `Lookup Value Code` field on the sales header, and a page control for this field on each of the sales document pages.

The next code snippet implements the extension of the `Sales Header` table, that is, scenarios #0004 and #0005:

```
tableextension 50001 "SalesHeaderTableExt" extends "Sales Header"
{
  fields
  {
    field(50000; "Lookup Value Code"; Code[10])
    {
      Caption = 'Lookup Value Code';
      DataClassification = ToBeClassified;
      TableRelation = "LookupValue";
    }
  }
}
```

Furthermore, the following code block will implement the extension of the `Sales Order` page (see scenario #0007):

```
pageextension 50002 "SalesOrderPageExt" extends "Sales Order"
{
  layout
  {
    addlast(General)
    {
      field("Lookup Value Code"; "Lookup Value Code")
      {
        ToolTip = 'Specifies the lookup value the transaction is
          done for.';
        ApplicationArea = All;
      }
    }
  }
}
```

Scenarios #0006, #0008, #0009, #0010, and #0011 would lead in a similar manner to the extension of the `Sales Quote`, `Sales Invoice`, `Sales Credit Memo`, `Sales Return Order`, and `Blanket Sales Order` document pages. Please refer to the GitHub repo to find all the code.

Test Code

With some big steps, you'll create your test code for scenarios #0004, #0006, and #0007, leaving scenarios #0005, #0008, #0009, #0010, and #0011 for you to review on GitHub.

Create a test codeunit

```
codeunit 81001 "LookupValue UT Sales Document"
{
  Subtype = Test;

  trigger OnRun()
  begin
    // [FEATURE] LookupValue UT Sales Document
  end;
}
```

Embed the customer wish into a test function

Embedding the three scenarios, #0004, #0006, and #0007, into test functions makes your new test codeunit look as follows:

```
codeunit 81001 "LookupValue UT Sales Document"
{
  Subtype = Test;

  trigger OnRun()
  begin
    //[FEATURE] LookupValue UT Sales Document
  end;

  [Test]
  procedure AssignLookupValueToSalesHeader()
  begin
    //[SCENARIO #0004] Assign lookup value to sales header page
    //[GIVEN] Lookup value
    //[GIVEN] Sales header
    //[WHEN] Set lookup value on sales header
    //[THEN] Sales header has lookup value code field populated
  end;

  [Test]
  procedure AssignLookupValueToSalesQuoteDocument()
  begin
    //[SCENARIO #0006] Assign lookup value on sales quote
    //                 document page
    //[GIVEN] Lookup value
    //[GIVEN] Sales quote document page
    //[WHEN] Set lookup value on sales quote document
    //[THEN] Sales quote has lookup value code field populated
  end;
```

```
  [Test]
  procedure AssignLookupValueToSalesOrderDocument()
  begin
    //[SCENARIO #0007] Assign lookup value on sales order
    //                      document page
    //[GIVEN] Lookup value
    //[GIVEN] Sales order document page
    //[WHEN] Set lookup value on sales order document
    //[THEN] Sales order has lookup value code field populated
  end;
}
```

Write your test story

Now that the structure is clear, we'll pick out scenario #0007 to create more detail:

```
codeunit 81001 "LookupValue UT Sales Document"
{
  Subtype = Test;

  trigger OnRun()
  begin
    //[FEATURE] LookupValue UT Sales Document
  end;

  [Test]
  procedure AssignLookupValueToSalesOrderDocument()
  begin
    //[SCENARIO #0007] Assign lookup value on sales order
    //                      document page

    //[GIVEN] Lookup value
    CreateLookupValueCode();
    //[GIVEN] Sales order document page
    CreateSalesOrderDocument();
    //[WHEN] Set lookup value on sales order document
    SetLookupValueOnSalesOrderDocument();
```

```
    //[THEN] Sales order has lookup value code field populated
    VerifyLookupValueOnSalesHeader();
  end;
}
```

So, how do you go about setting up the shared fixture? You do this by implementing the `Initialize` function instead of `CreateLookupValueCode`, as introduced in *Chapter 5*, *Test Plan and Test Design*. This would change `AssignLookupValueToSalesOrderDocument` into the following:

```
[Test]
procedure AssignLookupValueToSalesOrderDocument()
begin
  //[SCENARIO #0007] Assign lookup value on sales order
  //                 document page

  //[GIVEN] Lookup value
  Initialize();
  //[GIVEN] Sales order document page
  CreateSalesOrderDocument();
  //[WHEN] Set lookup value on sales order document
  SetLookupValueOnSalesOrderDocument();
  //[THEN] Sales order has lookup value code field populated
  VerifyLookupValueOnSalesHeader();
end;
```

Construct the real code

Let's build a simple `Initialize`:

```
local procedure Initialize()
begin
  if isInitialized then
    exit;

  //[GIVEN] Lookup value
  LookupValueCode := CreateLookupValueCode();
```

```
    isInitialized := true;
    Commit();
  end;
```

Here, both `isInitialized` and `LookupValueCode` are global variables of, respectively, the `Boolean` and `Code[10]` data types. Once `Initialize` has been called, `isInitialized` will be `true` and the `if` statement will evaluate to `true` the next time `Initialize` is being called, always leading to a straight exit from `Initialize`. This way, the Lookup Value record is inserted only once for this codeunit and reused as the global variable in all other test functions.

With respect to scenario #0007, your test codeunit would become as follows, including the various variables, parameters, and other helper functions:

```
codeunit 81001 "LookupValue UT Sales Document"
{
  Subtype = Test;

  trigger OnRun()
  begin
    //[FEATURE] LookupValue UT Sales Document
  end;

  var
    Assert: Codeunit "Library Assert";
    LibrarySales: Codeunit "Library - Sales";
    isInitialized: Boolean;
    LookupValueCode: Code[10];

  [Test]
  procedure AssignLookupValueToSalesOrderDocument()
  var
    SalesHeader: Record "Sales Header";
    SalesDocument: TestPage "Sales Order";
    DocumentNo: Code[20];
  begin
    //[SCENARIO #0007] Assign lookup value on sales order
    //                 document page
```

```
  //[GIVEN] Lookup value
  Initialize();
  //[GIVEN] Sales order document page
  CreateSalesOrderDocument(SalesDocument);
  //[WHEN] Set lookup value on sales order document
  DocumentNo := SetLookupValueOnSalesOrderDocument(
      SalesDocument, LookupValueCode);
  //[THEN] Sales order has lookup value code field populated
  VerifyLookupValueOnSalesHeader(
    "Sales Document Type"::Order, DocumentNo, LookupValueCode);
end;

local procedure Initialize()
begin
  if isInitialized then
    exit;

  //[GIVEN] Lookup value
  LookupValueCode := CreateLookupValueCode();

  isInitialized := true;
  Commit();
end;

local procedure CreateLookupValueCode(): Code[10]
begin
  //for implementation see test example 1; this smells
  //like duplication ;-)
end;

local procedure CreateSalesOrderDocument(
  var SalesDocument: TestPage "Sales Order")
begin
  SalesDocument.OpenNew();
end;
```

```
local procedure SetLookupValueOnSalesOrderDocument(
  var SalesDocument: TestPage "Sales Order";
  LookupValueCode: Code[10])
  DocumentNo: Code[20]
begin
  with SalesDocument do begin
    //for rest of implementation see test example 1
  end;
end;


local procedure VerifyLookupValueOnSalesHeader(
  DocumentType: Enum "Sales Document Type";
  DocumentNo: Code[20];
  LookupValueCode: Code[10])
var
  SalesHeader: Record "Sales Header";
  FieldOnTableTxt: Label '%1 on %2';
begin
  SalesHeader.Get(DocumentType, DocumentNo);
  //for rest of implementation see test example 1
  //another smell of duplication ;-)
end;
}
```

Test execution

Running a full-fledged `codeunit 81001` – including all the other scenarios to be found on GitHub – yields a bunch of successes – **green**! – as shown in *Figure 7.1*:

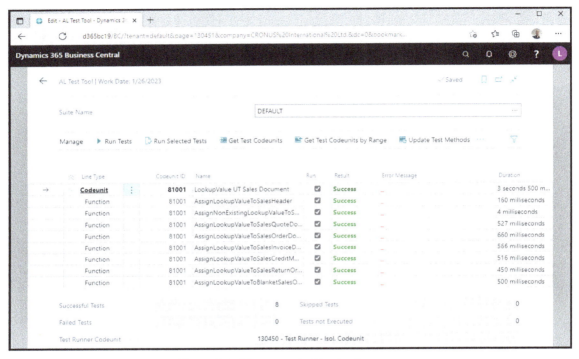

Figure 7.1 – Test example 4 having been run

Test the test

By now, I guess you know what to do here: adjust the test so the verification errs. Give it a try or use the completed code on GitHub as your cheat sheet.

Test example 5 – how to parameterize tests

Writing test automation, including design and coding, is a considerable effort, which has a lot of details to pay attention to. However, once you've got the hang of it and have it in place, you will enjoy it and continue to profit from it. This is the case unless you're sloppy on the details at both the design and coding levels, and thus have to keep fixing your tests.

Nevertheless, you will enjoy writing even more if you make your tests generic by **parameterizing** them. By the nature of the testability framework, you will not be able to parameterize a test function, but you can achieve this by encapsulating your generic test code in a helper function.

Customer wish

Let's illustrate this with another part of our customer's wish: archiving a sales document. As Business Central enables the user to archive a sales quote, a sales order, and a sales return order, we have to include this in our extension. This is expressed in the following three scenarios:

```
[FEATURE] LookupValue Sales Archive


[SCENARIO #0018] Archive sales order with lookup value
[GIVEN] Sales order with lookup value
[WHEN] Sales order is archived
[THEN] Archived sales order has lookup value from sales order


[SCENARIO #0019] Archive sales quote with lookup value
[GIVEN] Sales quote with lookup value
[WHEN] Sales quote is archived
[THEN] Archived sales quote has lookup value from sales quote


[SCENARIO #0020] Archive sales return order with lookup value
[GIVEN] Sales return order with lookup value
[WHEN] Sales return order is archived
[THEN] Archived sales return order has lookup value from sales
       return order
```

Note the GIVEN, which is a condensed version – a bigger step! – of the following tags in previous scenarios of *Test example 4 – how to set up a shared fixture*:

```
[GIVEN] Lookup value
[GIVEN] Sales order document page
[WHEN] Set lookup value on sales order document
```

Application code

The data model extension is implemented by the following .al object:

```
tableextension 50009 "SalesHeaderArchiveTableExt"
           extends "Sales Header Archive"
{
  fields
```

```
  {
    field(50000; "Lookup Value Code"; Code[10])
    {
      Caption = 'Lookup Value Code';
      DataClassification = ToBeClassified;
      TableRelation = "LookupValue";
    }
  }
}
```

And the UI is extended subsequently for scenario #0019 as per the code below. It will look very much alike for scenarios #0018 and #0020:

```
pageextension 50042 "SalesQuoteArchivePageExt"
                extends "Sales Quote Archive"
{
  layout
  {
    addlast(General)
    {
      field("Lookup Value Code"; "Lookup Value Code")
      {
        ToolTip = 'Specifies the lookup value the
          transaction is done for.';
        ApplicationArea = All;
      }
    }
  }
}


pageextension 50045 "SalesQuoteArchivesPageExt"
                extends "Sales Quote Archives"
{
  layout
  {
    addfirst(Control1)
    {
      field("Lookup Value Code"; "Lookup Value Code")
```

```
      {
        ToolTip = 'Specifies the lookup value the
          transaction is done for.';
        ApplicationArea = All;
      }
    }
  }
}
```

Test code

Now that the app code has been set, let's have a look at the test code.

Create, embed, and write

With one big step of *creating*, *embedding*, and *writing*, this is what the test stories #0018, #0019, and #0020 would look like when placed in a new test codeunit:

```
codeunit 81004 "LookupValue Sales Archive"
{
  Subtype = Test;

  trigger OnRun()
  begin
    //[FEATURE] LookupValue Sales Archive
  end;

  [Test]
  procedure ArchiveSalesOrderWithLookupValue()
  begin
    //[SCENARIO #0018] Archive sales order with lookup value
    //[GIVEN] Sales order with lookup value
    CreateSalesOrderWithLookupValue();
    //[WHEN] Sales order is archived
    ArchiveSalesOrderDocument();
    //[THEN] Archived sales order has lookup value from sales
    //       order
    VerifyLookupValueOnSalesOrderArchive();
```

```
end;

[Test]
procedure ArchiveSalesQuoteWithLookupValue()
begin
    //[SCENARIO #0019] Archive sales quote with lookup value
    //[GIVEN] Sales quote with lookup value
    CreateSalesQuoteWithLookupValue();
    //[WHEN] Sales quote is archived
    ArchiveQuoteDocument();
    //[THEN] Archived sales quote has lookup value from sales
    //         quote
    VerifyLookupValueOnSalesQuoteArchive();
end;

[Test]
procedure ArchiveSalesReturnOrderWithLookupValue()
begin
    //[SCENARIO #0020] Archive sales return order with lookup
    //                 value
    //[GIVEN] Sales return order with lookup value
    CreateSalesReturnOrderWithLookupValue();
    //[WHEN] Sales return order is archived
    ArchiveSalesReturnOrderDocument();
    //[THEN] Archived sales return order has lookup value from
    //       sales return order
    VerifyLookupValueOnSalesReturnOrderArchive();
end;
}
```

Construct the real code

When all three scenarios are testing the process of archiving a sales document, they boil down to a generic story with only the document type as a variable – quote, order, or return order. Consequently, you can consolidate this into one test story:

```
[Test]
procedure ArchiveSalesDocumentWithLookupValue()
begin
    //[SCENARIO #....] Archive sales document with lookup
    //                 value

    //[GIVEN] Sales document with lookup value
    CreateSalesDocumentWithLookupValue();
    //[WHEN] Sales document is archived
    ArchiveSalesDocument();
    //[THEN] Archived sales document has lookup value from
    //       sales document
    VerifyLookupValueOnSalesDocumentArchive();
end;
```

The only problem is that you cannot parameterize a `test` function. To solve this situation, you can create a local method to be called from the three tests:

```
local procedure ArchiveSalesDocumentWithLookupValue(
    DocumentType: Enum "Sales Document Type"): Code[20]
var
    SalesHeader: record "Sales Header";
begin
    //[GIVEN] Sales document with lookup value
    CreateSalesDocumentWithLookupValue(SalesHeader, DocumentType);
    //[WHEN] Sales document is archived
    ArchiveSalesDocument(SalesHeader);
    //[THEN] Archived sales document has lookup value from sales
    //       document
    VerifyLookupValueOnSalesDocumentArchive(
        DocumentType,
        SalesHeader."No.",
        SalesHeader."Lookup Value Code",
```

```
    1);   // Used 1 for No. of Archived Versions
  exit(SalesHeader."No.")
end;
```

The three tests will then become:

```
[Test]
procedure ArchiveSalesOrderWithLookupValue()
begin
  //[SCENARIO #0018] Archive sales order with lookup value
  ArchiveSalesDocumentWithLookupValue(
    "Sales Document Type"::Order)
end;


[Test]
procedure ArchiveSalesQuoteWithLookupValue()
begin
  //[SCENARIO #0019] Archive sales quote with lookup value
  ArchiveSalesDocumentWithLookupValue(
    "Sales Document Type"::Quote)
end;


[Test]
procedure ArchiveSalesReturnOrderWithLookupValue()
begin
  //[SCENARIO #0020] Archive sales return order with lookup
  //                 value
  ArchiveSalesDocumentWithLookupValue(
    "Sales Document Type"::"Return Order")
end;
```

This is easy to achieve with **copy** and **paste**: three birds with one stone.

Note

Go to GitHub to have a look at the implementation of the other helper functions and an additional scenario, #00021. Note that the usage of the enqueue and dequeue functions will be discussed in the next test example.

Test execution

Show me the **green** successes:

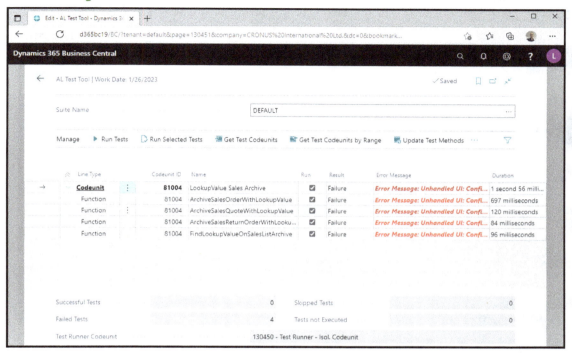

Figure 7.2 – Test example 5 having been run

Ouch ...... **red. RED**?

Apparently, as the test errors indicate in Test Tool, we need to handle a Confirm. You might want to click on the hyperlinked error to show the full text in the overlay dialog:

```
Error Message: Unhandled UI: Confirm Archive Order no.: 1001?
```

Before you continue with your tests, let's go into the application and try to archive a sales order. To accomplish this, take the following steps:

1. Use **Alt + Q** to trigger the **Tell Me What You Want** feature.
2. Type Sales Orders and select the **Sales Orders** hyperlink to open the Sales Orders page.
3. Open the document page of the first sales order.

4. Select **Actions | Functions | Archive Document**.

    Indeed, a dialog appears here asking the user to confirm (or not):

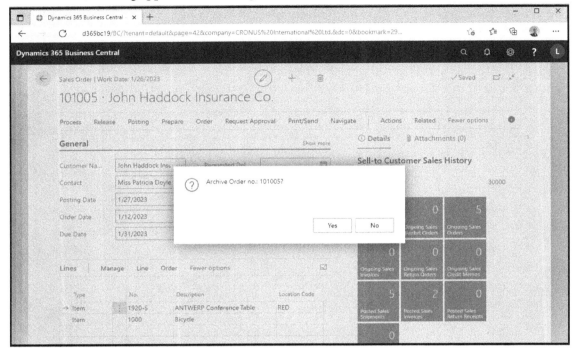

Figure 7.3 – Archive Order confirm dialog

5. See what happens when we click **Yes** in the confirm dialog: a message appears informing the user that the document has been archived, as shown in *Figure 7.4*:

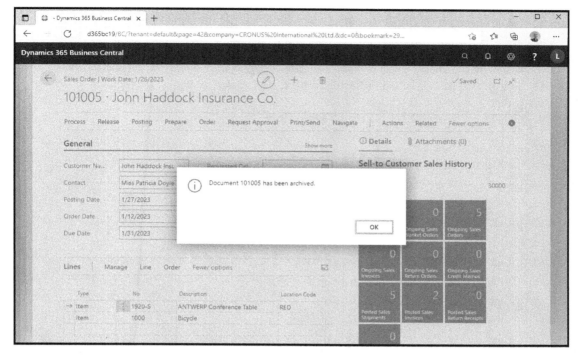

Figure 7.4 – Document archived message

Once you click **OK** in the message dialog, the archiving of the document is complete.

With respect to your test automation, you need to create two handler functions – one handler function to handle the confirm dialog, and the other to handle the message, as follows:

```
[ConfirmHandler]
procedure ConfirmHandlerYes(Question: Text[1024]; var Reply:
Boolean);
begin
  Reply := true;
end;


[MessageHandler]
procedure MessageHandler(Message: Text[1024]);
begin
end;
```

As mentioned in *Chapter 3*, *The Testability Framework*, just defining the handlers is not sufficient. They also need to be linked to the test(s), triggering the UI elements, using the `HandlerFunctions` tag:

```
[HandlerFunctions('ConfirmHandlerYes,MessageHandler')]
```

The test codeunit for scenario #0018 will then become:

```
[Test]
[HandlerFunctions('ConfirmHandlerYes,MessageHandler')]
procedure ArchiveSalesOrderWithLookupValue()
var
  SalesHeader: record "Sales Header";
begin
  //[SCENARIO #0018] Archive sales order with lookup value
  ArchiveSalesDocumentWithLookupValue(
    "Sales Document Type"::Order)
end;
```

Now, run it again! Do show us the **green**, please:

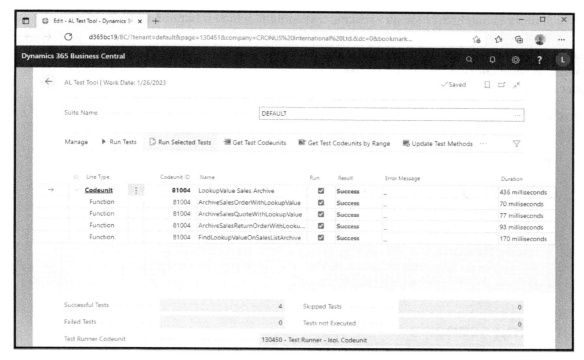

Figure 7.5 – Test example 5 having been run successfully

> **Note**
> Waldo's CRS AL Language Extension contains for each UI handler a code snippet with the right signature.

Test the test

You know what to do. Yes, you do, right?

A note on verifying the handlers

Both the handlers are now minimally implemented. They will handle the dialogs, but not check anything. In other words, these handlers will simply close the dialog, regardless of the content. In real life, this might also be a very different message or confirm dialog than we had expected, and you will need to program to interact with the dialog accordingly.

As each message/confirm has a very specific text content, which is handed off to the handler by the system as the `Question`/`Message` parameter, you will have to write additional code in the handler to verify whether it is triggered by the right dialog. If so, continue – and let the handler mimic the user clicking the **OK** or **Yes/No** button – if not, throw an error. This way, you make your handler robust.

Have a look at the final code in the GitHub repo. The principle used to hand over (parts of) the expected question/message will be discussed in the next test example.

> **Note**
> A test function can only have one handler function of a type linked to it. So, if your test does trigger multiple UI elements of the same type, these will need to be handled by only one UI handler. In case multiple messages pop up during the execution of one test, these will all need to be handled by one Message Handler. If you happen to link multiple handlers of the same type to a test function, only the first one listed will be used for the associated UI elements. Your test will fail, and you will get an error message that the other handler did not run.

A missing scenario?

One of the great reviewers of the first edition of this book, Steven Renders, pointed out to me that there is a hole in the scenarios for the customer wish, which is, when archiving a sales document, the lookup value should be carried over to the archived sales document. Before I step into the specifics, this is a perfect illustration of what I already mentioned in *Chapter 5, Test Plan and Test Design*:

"A test design is an object to help the team discuss their test effort, to reveal the holes in their thoughts…"

So, what is this hole? If you have a confirmation, asking the user for a **Yes** or **No**, there are potentially two scenarios that need to be tested, and my scenarios only handle the **Yes**. So, what about the **No**? This indeed is a user scenario as such, but I do not consider it a scenario to be tested within the context of our customer wish, because when the user clicks **No**, the transaction does not complete. It is a scenario that relates to the bigger feature of archiving sales documents. As such, this scenario was not added to our collection on the assumption that this is handled by standard tests. You could argue whether you want to include a test for this anyway, but you now know how to program a handler for that scenario ☺.

Nevertheless, in any future project, be triggered when confirm statements are used, as in principle, these can potentially lead to two scenarios.

Test example 6 – how to hand over data to UI handlers

Just now, with the previous test example where we hit upon the need for two dialog handlers, it makes sense to discuss how to hand over data to a **UI handler**, as we cannot directly control this. It's the platform that controls this, and we have no way of passing any additional data.

Customer wish

In this context, let's pick up another part of our customer wish – when creating a new customer from the UI, by clicking the standard **New** action on the ribbon, you have to select a template to base the new customer on (or simply bypass it by selecting **Cancel**), as shown in *Figure 7.6*:

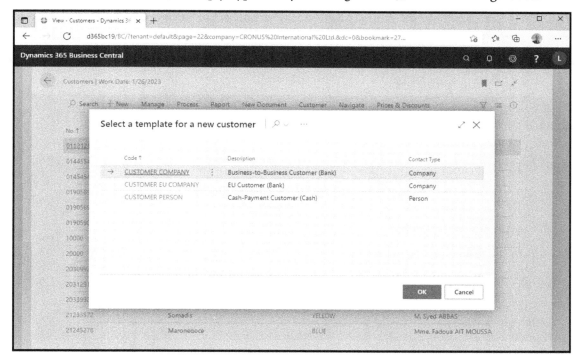

Figure 7.6 – Select a template for a new customer

This part of the customer wish tells you that the configuration template that's behind the templates you can choose should be set up so that it will auto-populate the `Lookup Value Code` field on the newly created customer from the selected template.

> **Notes**
>
> (1) We've had to tackle the appearance of the ModalPage already in test example 3 of the previous chapter.
>
> (2) The customer template functionality has changed in BC19. The GitHub repo tests for both BC19 and BC18 are present in the *main* and *BC18* branches, respectively.

This is what scenario #0028 entails:

```
[FEATURE] LookupValue Inheritance
[SCENARIO #0028] Create customer from configuration
                 template with lookup value
[GIVEN] Configuration template (customer) with lookup value
[WHEN] Create customer from configuration template
[THEN] Lookup value on customer is populated with lookup
       value of configuration template
```

We can accomplish this by setting up a configuration template. There is no need for any additional application code.

Test code

Let's wrap scenario #0028 in a new test codeunit.

Create, embed, and write

This leads to the following code construction:

```
codeunit 81006 "LookupValue Inheritance"
{
    Subtype = Test;

    trigger OnRun()
    begin
        //[FEATURE] LookupValue Sales Document / Customer
    end;
```

```
  [Test]
  procedure
    InheritLookupValueFromConfigurationTemplateToCustomer()
  begin
    //[SCENARIO #0028] Create customer from configuration
    //                 template with lookup value
    Initialize();

    //[GIVEN] Configuration template (customer) with lookup
    //        value
    CreateCustomerConfigurationTemplateWithLookupValue();
    //[WHEN] Create customer from configuration template
    CreateCustomerFromConfigurationTemplate();
    //[THEN] Lookup value on customer is populated with lookup
    //       value of configuration template
    VerifyLookupValueOnCustomer();
  end;
}
```

Construct the real code

Including all technical details, like variables and arguments, this codeunit would become:

```
codeunit 81006 "LookupValue Inheritance"
{
  Subtype = Test;

  trigger OnRun()
  begin
    //[FEATURE] LookupValue Sales Document / Customer
  end;

  [Test]
  [HandlerFunctions('HandleConfigTemplates')]
  procedure
    InheritLookupValueFromConfigurationTemplateToCustomer()
  var
    CustomerNo: Code[20];
```

```
      ConfigTemplateHeaderCode: Code[10];
  begin
    //[SCENARIO #0028] Create customer from configuration
    //                 template with lookup value
    Initialize();

    //[GIVEN] Configuration template (customer) with lookup
    //        value
    ConfigTemplateHeaderCode :=
      CreateCustomerConfigurationTemplateWithLookupValue(
        LookupValueCode);
    //[WHEN] Create customer from configuration template
    CustomerNo :=
      CreateCustomerFromConfigurationTemplate(
        ConfigTemplateHeaderCode);
    //[THEN] Lookup value on customer is populated with lookup
    //        value of configuration template
    VerifyLookupValueOnCustomer(CustomerNo,
      LookupValueCode);
  end;
}
```

We need to create the following four helper functions and one UI handler:

- `Initialize`
- `CreateCustomerConfigurationTemplateWithLookupValue`
- `CreateCustomerFromConfigurationTemplate`
- `VerifyLookupValueOnCustomer`
- `HandleConfigTemplates`

Two of the five procedures needed can be inherited from earlier test examples:

- `Initialize` takes care of the Lookup Value and can be copied from test example 4.
- `VerifyLookupValueOnCustomer` can be taken from test example 1.

The other three functions,
`CreateCustomerConfigurationTemplateWithLookupValue`,
`CreateCustomerFromConfigurationTemplate`, and `HandleConfigTemplates`,
will be as the following code block.

The function names describe exactly what the function is doing. I will leave it to you to read and
grasp the meaning of the first two. In the context of this test example, we will elaborate more on
`HandleConfigTemplates`:

```
local procedure CreateCustomerConfigurationTemplateWithLookupValue(
    LookupValueCode: Code[10]): Code[10]
// Adopted from Codeunit 132213 Library - Small Business
var
  ConfigTemplateHeader: record "Config. Template Header";
  Customer: Record Customer;
begin
  LibraryRapidStart.CreateConfigTemplateHeader(
    ConfigTemplateHeader);
  ConfigTemplateHeader.Validate("Table ID", Database::Customer);
  ConfigTemplateHeader.Modify(true);

  LibrarySmallBusiness.CreateCustomerTemplateLine(
    ConfigTemplateHeader,
    Customer.FieldNo("Lookup Value Code"),
    Customer.FieldName("Lookup Value Code"), LookupValueCode);

  exit(ConfigTemplateHeader.Code);
end;


local procedure CreateCustomerFromConfigurationTemplate(
    ConfigurationTemplateCode: Code[10]) CustomerNo:  Code[20]
var
  CustomerCard: TestPage "Customer Card";
begin
  CustomerCard.OpenNew();
  CustomerNo := CustomerCard."No.".Value();
  CustomerCard.Close();
end;
```

```
[ModalPageHandler]
procedure HandleConfigTemplates(
    var ConfigTemplates: TestPage "Config Templates")
begin
  ConfigTemplates.GoToKey(
    <provide the PK of the Config Template>);
  ConfigTemplates.OK.Invoke();
end;
```

As soon as a new customer card is created in
CreateCustomerFromConfigurationTemplate, the Config Templates page
needs to be handled by the ModalPageHandler HandleConfigTemplates. Out of
the list of configuration templates, it should select the configuration template created by
CreateCustomerConfigurationTemplateWithLookupValue.

With the GoToKey method of a TestPage, we can achieve this, but it needs to provide the PK
value of the template, as marked by the triangular brackets in the preceding code.

A straightforward solution would be to create a global variable called ConfigTemplateCode,
which would be populated in the **GIVEN** part of our test, as follows:

```
ConfigTemplateCode :=
  CreateCustomerConfigurationTemplateWithLookupValue(
    LookupValueCode);
```

This would successively be picked up by our ModalPageHandler. This would undoubtedly
be a perfectly valid solution. But picture yourself having to hand over multiple different values
of different data types in one test codeunit, stacking global variable upon global variable. To
overcome this, Microsoft has provided us with a neat feature implemented in the codeunit,
Library - Variable Storage. It consists of a queue of 25 variant elements. Using Enqueue and
Dequeue, you can store and retrieve your variable in a first in, first out manner. Be sure to have
an Enqueue balanced by an Dequeue.

Enqueue

Call Enqueue in your test code before the handler is triggered:

```
//[GIVEN] Configuration template (customer) with lookup value
ConfigTemplateCode :=
  CreateCustomerConfigurationTemplateWithLookupValue(
    LookupValueCode);
//[WHEN] Create customer from configuration template
```

```
LibraryVariableStorage.Enqueue(ConfigTemplateCode);
CustomerNo :=
  CreateCustomerFromConfigurationTemplate(
    ConfigTemplateCode);
```

Dequeue

In the handler, call Dequeue to retrieve the variable:

```
[ModalPageHandler]
procedure HandleConfigTemplates(
    var ConfigTemplates: TestPage "Config Templates")
begin
  ConfigTemplates.GoToKey(LibraryVariableStorage.
    DequeueText());
  ConfigTemplates.OK.Invoke();
end;
```

Test execution

Fingers crossed for **green** results:

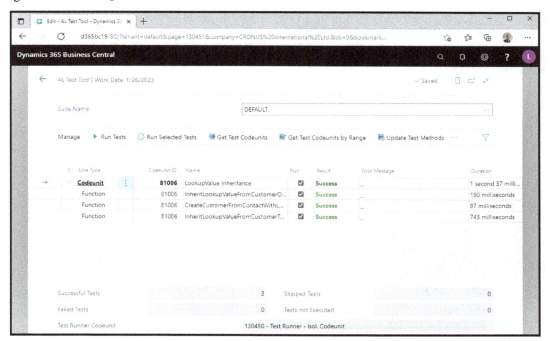

Figure 7.7 – Test example 6 having been run

Success!

> **Note**
>
> Notice the two other test function lines of codeunit `81006`, `LookupValue`
> `Inheritance`, containing two other scenarios, `#0024` and `#0026`. You will find
> their details – the ATDD scenario and test function – on GitHub.

Test the test

By now, you know how to adjust a test, so the verification errs. But does the queue do its job right? How about enqueuing a non-existent configuration template code? Let's randomly choose one – `LUC`.

Running the test now throws the following error:

```
Unexpected CLR exception thrown.: Microsoft.Dynamics.
Framework.UI.FormAbortException: Page New - Customer Card
has to close ---> Microsoft.Dynamics.Nav.Types.Exceptions.
NavTestRowNotFoundException: The row does not exist on the
TestPage. ---> System.
```

Figure 7.8 shows the error on screen:

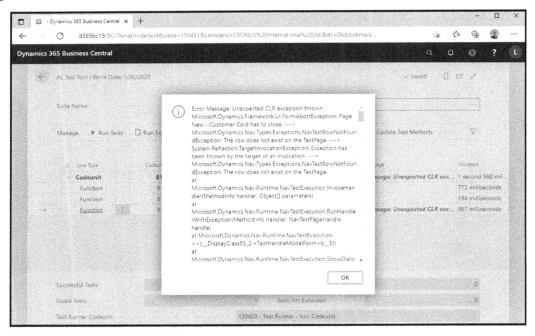

Figure 7.8 – Testing test example 6 throwing an error

The error message does not mention the row key value, but it surely lets us know that it cannot find the row the test wants to be selected, that is, LUC.

Now, with this test example, and the two previous ones, you've added another set of tools to your test automation toolkit.

Summary

By building three more test examples in this chapter, you learned how to set up a shared fixture, how to parameterize tests, and how to hand over variables to UI handlers. These are three techniques that will be of invaluable worth in your future test automation practices.

In the next chapter, we will add two more *tools* to your test toolkit. You will learn how to test a report dataset and how to work out a more complex scenario.

8

From Customer Wish to Test Automation – the TDD way

Getting the hang of it by now? Well, you did get this far, right? How about doing it **TDD** way now, putting *test first* all the way? And while doing that I'll supplement your toolbox with some more *test tools* for Microsoft Dynamics 365 Business Central. In this chapter, we'll expand on how to:

- Refactor your code (*Test example 7*)
- Test a report (*Test example 8*)
- Test with permissions (*Test example 9*)

But before we do, let's review the previous test examples in the TDD light.

Technical requirements

Like in the previous chapter we will refer to the **LookupValue** extension. Its code can be found on GitHub: `https://github.com/PacktPublishing/Automated-Testing-in-Microsoft-Dynamics-365-Business-Central-Second-Edition`.

In this chapter, we will continue on the code from *Chapter 7, From Customer Wish to Test Automation – Next Level*: `https://github.com/PacktPublishing/Automated-Testing-in-Microsoft-Dynamics-365-Business-Central-Second-Edition/tree/main/Chapter 07 (LookupValue Extension)`.

The final code of this chapter can be found in `https://github.com/PacktPublishing/Automated-Testing-in-Microsoft-Dynamics-365-Business-Central-Second-Edition/tree/main/Chapter 08 (LookupValue Extension)`.

Details on how to use this repository and how to set up VS Code are discussed in *Appendix, Getting Up and Running with Business Central, VS Code, and the GitHub Project*.

TDD and our test examples

After I introduced the concept of **Test-Driven Development** in *Chapter 2, Test Automation and Test-Driven Development*, I did not pay any attention to the topic until now. In fact, I deliberately evaded to use TDD as our way of working for mainly one reason:

To write automated tests for application code that already exists.

This is most probably the context in which most of you will start applying test automation.

Now, what if you had applied TDD to our test examples?

To be honest, it wouldn't have looked much different, as a lot of the TDD principles were implicitly exercised by:

- Defining your customer wishes by means of the **ATDD** scenarios, you created yourselves a sufficient set of tests, that is, a test list.
- Implementing your tests with the 4-steps recipe.

With the latter, we:

- Took small steps.
- Created a structure for each test based on the `GIVEN`-`WHEN`-`THEN` tags.
- Constructed the real code to get it to work.
- Ran the test, and if `red`, adjusted the test code until the test passed, that is, `green`.

The only deviations from TDD were that the application code was conceived before you started working on the test code, the app code functioned right, and the fact that you did not refactor anything. Sounds like it wouldn't be too big a step to start exercising TDD in your daily work, doesn't it? To sing the **red-green-refactor** mantra as introduced in *Chapter 2, Test Automation and Test-Driven Development*, you'll have to:

1. Take a test from the test list and write the test code.
2. Compile the test code yielding red as the application code is not yet there.
3. Implement *just enough* application code to make the test code compile.
4. Run the test seeing it probably fail, still red.
5. Adjust the application code *just enough* to make it pass, that is, green.
6. **Refactor** your code (either test or application code, or both), one after the other, and rerun the test after each change to prove all code is still well (green).
7. Move to the next test on the list and repeat from **Step 1**.

As discussed in *Chapter 2* you can perfectly use TDD for Business Central. In my experience, the biggest challenge is to create the discipline in your team to spec the customer wish in scenarios.

In this chapter, in *Test examples 8* and *9*, we will explicitly reference the foregoing steps instead of the *4-steps recipe* even though the latter will still be our implicit guide in how to convert an ATDD scenario into test code. But before we do, we will not start at **Step 1** of the **red-green-refactor** mantra but instead *wipe the slate clean* to make the *singing* easier. Thus, we will solve our **refactor** backlog first and start with **Step 6** of the mantra in *Test example 7*.

Test example 7 – how to refactor your code

From a TDD perspective, the previous six test examples fully neglected a very essential part: the *refactoring of the code*, or more accurately, one of the two rules that make TDD: **Rule 2 Eliminate duplication**.

As I hope you will experience here, refactoring is not rocket science but a discipline, as I tend to call it. **Refactoring** is not a goal on its own, but a means to exercise this second rule of TDD, with the valuable resulting effects of getting **reusable**, **readable**, and **minimalistic** code.

Contrary to what many of us have been taught when becoming a developer, though, with TDD we do not design reusable, readable, and minimalistic code upfront. No, when performed with common coding sense, rule 2 leads you there. As you might recall from *Chapter 2, Test Automation and Test-Driven Development*, with TDD your first gear is taking small steps to be *effective and efficient*. Meaning that you don't do more than required for that step and don't be tempted to start generalizing your code if it's not needed to get it working.

So, here we go. Let's inspect the test code we have been implementing so far and see if we **smell** any duplication. Well, that shouldn't be too hard, as I did mark some code parts *smelling of duplication*, as you might recall.

> **Note**
> As the first focus with TDD is to get your code working and not to eliminate duplication I do often mark code as smelling to help remind me later. Of course, this does not say there are no other candidates for **duplication elimination**.

Smell of duplication 1

This was the first code snippet that I marked having a smell of duplication:

```
local procedure CreateLookupValueCode(): Code[10]
begin
  // for implementation see test example 1; this smells like
  // duplication ;-)
end;
```

In all of our test codeunits so far, implementing our ATDD features, this method has been coded in exactly the same way:

```
local procedure CreateLookupValueCode(): Code[10]
var
  LookupValue: Record LookupValue;
begin
  LookupValue.Init();
  LookupValue.Validate(
    Code,
    LibraryUtility.GenerateRandomCode(LookupValue.FieldNo(Code),
    Database::LookupValue));
  LookupValue.Validate(Description, LookupValue.Code);
  LookupValue.Insert();
  exit(LookupValue.Code);
end;
```

It could almost only smell less of duplication than this. So, let's eliminate the duplication by abstracting this method into a library to go with our tests:

```
codeunit 80000 "Library - Lookup Value"
{
  var
    LibraryUtility: Codeunit "Library - Utility";

  procedure CreateLookupValueCode(): Code[10]
  var
    LookupValue: Record LookupValue;
  begin
    LookupValue.Init();
    LookupValue.Validate(
      Code,
      LibraryUtility.GenerateRandomCode(
        LookupValue.FieldNo(Code),
        Database::LookupValue));
    LookupValue.Validate(Description, LookupValue.Code);
    LookupValue.Insert();
    exit(LookupValue.Code);
  end;
}
```

Now we can reference this library method in the local helper function by the same name in each of the test codeunits constructed so far:

```
local procedure CreateLookupValueCode(): Code[10]
begin
  exit(LibraryLookupValue.CreateLookupValueCode())
end;
```

> **Note**
> LibraryLookupValue is declared as a global variable based on this new codeunit, Library - Lookup Value.

Did we take a small enough step that our tests still run fine? Let's find out by rerunning all the tests we built so far: and **green**!

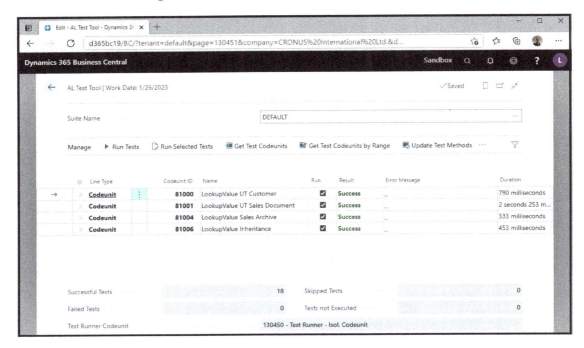

Figure 8.1 – Test Example 7 run after first refactoring: GREEN

Indeed, we could well digest the first refactoring step and our application and test code are cleaned up and still in balance. Maybe superfluous to say, but don't let yourself be pushed by the size of the step we're taking here. If a smaller step fits you better, do not hesitate to apply it. A smaller step in this context could have been to rerun the tests after updating the local helper method `CreateLookupValueCode` in only one of the four test codeunits.

> **Note**
>
> A smaller step would be to not redirect each local helper method `CreateLookupValueCode` to the new library function at the same time. Instead, do that one by one and rerun your tests after each redirect. This is what is called parallel change, which is commonly used in refactors to reduce risk. Read more at `https://www.martinfowler.com/bliki/ParallelChange.html`.

Smell of duplication 2

Another smell clearly marked is the following:

```
local procedure VerifyLookupValueOnSalesHeader(
  DocumentType: Enum "Sales Document Type";
  DocumentNo: Code[20];
  LookupValueCode: Code[10])
var
  SalesHeader: Record "Sales Header";
  FieldOnTableTxt: Label '%1 on %2';
begin
  SalesHeader.Get(DocumentType, DocumentNo);
  // for rest of implementation see test example 1 another smell
  // of duplication ;-)
end;
```

The duplication here concerns the construction of the additional message provided to the `AreEqual(Expected: Variant, Actual: Variant, Msg: Text)` method based on the `FieldOnTableTxt` label variable and highlighted in the following code snippet:

```
Assert.AreEqual(
  LookupValueCode,
  SalesHeader."Lookup Value Code",
  StrSubstNo(
    FieldOnTableTxt,
    SalesHeader.FieldCaption("Lookup Value Code"),
    SalesHeader.TableCaption())
  );
```

This construction indeed is duplicated in all five test codeunits created so far, and I have abstracted it into a new library:

```
codeunit 80001 "Library - Messages"
{
  procedure GetFieldOnTableTxt(
    FieldCaption: Text; TableCaption: Text): Text
  var
    FieldOnTableTxt: Label '%1 on %2';
  begin
```

```
    exit(StrSubstNo(
      FieldOnTableTxt,
      FieldCaption,
      TableCaption))
  end;
}
```

The `AreEqual` call will then become:

```
Assert.AreEqual(
  LookupValueCode,
  SalesHeader."Lookup Value Code",
  LibraryMessages.GetFieldOnTableTxt(
    SalesHeader.FieldCaption("Lookup Value Code"),
    SalesHeader.TableCaption())
  );
```

Is our code still in balance? Will our tests still run fine? Run, run, run the tests again: **green**!

If your step was too big this could lead to a **red** test run result as you introduce more changes, making it more difficult to figure what caused the **red** result. The most obvious thing to do then is to revert your changes since the last **green**. If you did put your code under source code management, like I am doing right now with the GitHub repo going with this book, this should be a fairly easy thing to do.

> **Note**
> You might even practice so-called micro commits where you commit every time you *sing* the next step of the **red-green-refactor** mantra. Going to that last successful step is as fast as a rollback, nothing else to do. See also `https://www.industriallogic.com/blog/whats-this-about-micro-commits/`.

Some more duplications?

Where I had clearly marked the previous two smells, it is a little bit more challenging to find any other duplication, as we need to investigate the four test codeunits in more detail. This clearly shows the value of exercising **Step 6** straight after implementing a test from the test list.

You might want to test yourself by going through the final version of the code of *Chapter 7, From Customer Wish to Test Automation – Next Level*, and see what you would mark as duplication. I marked the following two, and eliminated them:

- `ValueCannotBeFoundInTableTxt`

  Our inspection will show that, similar to the second *smell*, another duplicate message construction can be found in codeunits `LookupValue UT Customer` (81000) and `LookupValue UT Sales Document` (81001) using the label `ValueCannotBeFoundInTableTxt`.

  You can check out yourself on GitHub how this duplication has been eliminated by adding the `GetValueCannotBeFoundInTableTxt` method to `Library - Messages` (80001).

- `VerifyLookupValueOnCustomer`

  The verification helper function `VerifyLookupValueOnCustomer`, first conceived in codeunit `LookupValue UT Customer` (81000), has an exact twin in codeunit `LookupValue Inheritance` (81006).

  You can check out yourself on GitHub how this duplication has been eliminated by adding the `VerifyLookupValueOnCustomer` method to `Library - Lookup Value` (80000).

> **Note**
>
> Indeed, you might have noticed that there are a few more code snippets/methods that have some smell of duplication. As these typically are not exact duplications, I, therefore, tend to leave them to evaluate them later once more, thus preventing some *made-up* generalization. As such I am practicing what is called the **rule of three**: https://en.wikipedia.org/wiki/Rule_of_three_(computer_programming).

Now that you have solved your refactoring backlog, you're fully set to go down the TDD road for your next test examples. Bear in mind that having a complete set of application and test code in place allows you to refactor whatever you need to refactor, be it application or test code. But be sure to do only one or the other. If refactoring application code leads to failing tests that used to succeed before, improve your refactored code and make all tests pass. If refactoring test code makes them fail where they didn't previously, revert and do it better.

Refactoring application code, not covered by tests yet

In case you need to refactor application code that is not yet covered by tests, write tests before starting. If you don't, the chances of breaking something and not noticing are very high.

Georgios Papoutakis of the Business Central development team suggests getting the test coverage of your code above 80%. Use the **Code Coverage** tool to measure the current coverage and add more tests until you hit the target. You could choose one of the two following strategies for adding tests depending on the investment you want to put into the project:

- **Option A**: Consider current code as the source of truth and write tests that verify the new code will behave as the old.

  This is the less costly strategy of the two.

- **Option B**: Define requirements and convert them to tests. Doing this, you will probably find issues with the current code as it does not – fully – comply with these requirements. Write tests that verify that the new code behaves as good as the old, but most likely even better.

  This is the most expensive of the two strategies.

> **Note**
>
> See *Chapter 9, How to Integrate Test Automation in Daily Development Practice*, for a discussion on the Code Coverage tool and the meaning of its measurement.

Test example 8 – how to test a report

Reports have always been a substantial part of many Business Central projects and solutions. It makes perfect sense to have a look at how to test them in an automated manner. But how do you go about doing that? In this example, you will learn how to test the dataset being created by a report by inspecting its XML structure. Layout testing is another job to do, and, unfortunately, one outside of the testability framework.

Customer wish

Your customer's wish describes that the `Lookup Value Code` field of the customer must be carried over to the various sales documents. The logical consequence, even though not explicitly stated, would be that each printed version of these documents would have to be extended with this field. As sales document reports are quite comprehensive, both on the dataset and on the layout side, we take a simpler example.

We'll clone report 101, `Customer - List`, and add the `Lookup Value Code` field to it as shown in *Figure 8.2*:

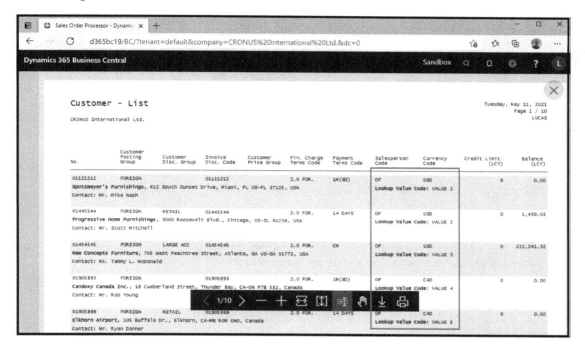

Figure 8.2 – Customer – List report extended with the Lookup Value Code field

But, let's remind ourselves that we're going down the TDD road. *Test first*, so, **Step 1 – take a test from the test list and write the test code**.

> **Note**
>
> With Business Central 2021 Wave 1 (BC18) we can extend a standard report using the newly introduced `reportextension` object. For the TDD flow, however, we have chosen to implement the customer wish on a cloned version of the standard report 101, `Customer - List`.

Step 1 – Take a test from the test list and write the test code

OK. What's on our **test list** with respect to the report?

Scenario #0029:

[FEATURE] LookupValue Report

[SCENARIO #0029] Test that lookup value shows on CustomerList
                 report

[GIVEN] 2 customers with different lookup value

[WHEN] Run report CustomerList

[THEN] Report dataset contains both customers with lookup value

> **Note**
>
> Why two customers, you might wonder? As the report should be able to list multiple customers, it makes sense not to test for just one customer.

And what should be the test code that goes with it? Have a look at our .al implementation of scenario #0029, where at first we *created* the codeunit, *embedded* the scenario, and *wrote* the story with the following result:

```
codeunit 81008 "Lookup Value Report"
{
  Subtype = Test;

  //[FEATURE] LookupValue Report

  [Test]
  procedure TestLookupValueShowsOnCustomerListReport();
  var
    Customer: array[2] of Record Customer;
  begin
    //[SCENARIO #0029] Test that lookup value shows on
    //                 CustomerList report

    //[GIVEN] 2 customers with different lookup value
    CreateCustomerWithLookupValue(Customer[1]);
    CreateCustomerWithLookupValue(Customer[2]);
```

```
// [WHEN] Run report CustomerList
CommitAndRunReportCustomerList();

// [THEN] Report dataset contains both customers with
//        value
VerifyCustWithLookupValueOnCustListReport(
    Customer[1]."No.", Customer[1]."Lookup Value Code");
VerifyCustWithLookupValueOnCustListReport(
    Customer[2]."No.", Customer[2]."Lookup Value Code");
    end;
}
```

Given this test function implementation, the helper functions must be constructed – *construct the real code* from the *4-steps recipe*. This is what they look like:

```
local procedure CreateCustomerWithLookupValue(
    var Customer: Record Customer)
begin
    LibrarySales.CreateCustomer(Customer);
    Customer.Validate("Lookup Value Code",CreateLookupValueCode());
    Customer.Modify();
    // have a look at scenario #0024 mentioned with test example 6
    // this smells like duplication ;-)
end;

local procedure CreateLookupValueCode(): Code[10]
var
    LookupValue: Record LookupValue;
begin
    // for implementation see test example 1; this also smells
    // like duplication ;-)
end;
```

`CreateCustomerWithLookupValue` and `CreateLookupValueCode` are becoming a kind of next-door neighbor, playing a role in most of our previous test examples. They clearly carry a *smell of code duplication*, making them candidates for refactoring.

> **Note**
> You might notice that global variables declaration has been omitted. Have a look at GitHub to see them being declared.

Testing a report dataset is about inspecting its XML structure, therefore, `CommitAndRunReportCustomerList` calls `RunReportAndLoad` in codeunit `Library - Report Dataset (131007)` to stream the dataset in a temporary `TempBlob` record (table `99008535`) to be used in the verification part:

```
local procedure CommitAndRunReportCustomerList()
var
  CustomerListReport: Report CustomerList;
  RequestPageXML: Text;
begin
  Commit(); // close open write transaction to be able to
            // run the report
  RequestPageXML := Report.RunRequestPage(
    Report::CustomerList,
    RequestPageXML);
  LibraryReportDataset.RunReportAndLoad(
    Report::CustomerList,
    '',
    RequestPageXML);
end;
```

Step 2 – Compile test code yielding red as application code is not yet there

If you are coding as you follow along with this example, you will observe that the reference to the report object, that is `Report::CustomerList`, is not recognized. It is VS Code executing **Step 2** of the *red-green-refactor* mantra telling us the object does not exist yet, thus our test fails: red!

Step 3 – Add just enough application code

Time to execute **Step 3 – implement just enough application code to make the test code compile**, and clone the standard report 101, Customer – List; both the report objects and its layout, which I have named CustomerList with ID 50000. I'll leave that to your AL development skills to get that done or call the GitHub repo to the rescue.

The test is almost ready. The fresh fixture has been implemented and the report is being exercised. This leaves us with the verification part. In VerifyCustWithLookupValueOnCustListReport, we make use of the AssertElementWithValueExists of Library - Report Dataset to verify whether a combination of an element name and value can be found in the XML structure:

- Element Customer_No_ with value No

- Element Customer_Lookup_Value_Code with value LookupValueCode

```
local procedure VerifyCustWithLookupValueOnCustListReport(
  No: Code[20]; LookupValueCode: Code[10])
begin
  LibraryReportDataset.AssertElementWithValueExists(
    'Customer_No_', No);
  LibraryReportDataset.AssertElementWithValueExists(
    'Customer_Lookup_Value_Code', LookupValueCode);
end;
```

Step 4 – Run test seeing it probably fail

Code compiles: time to deploy and run the test. Almost obviously expecting **green**. But let me remind you that we're in TDD mode, executing **Step 4** and not having implemented the `Lookup Value Code` column on the report, yet. The result has to be **red**. It would be very awkward if it turned out **green**.

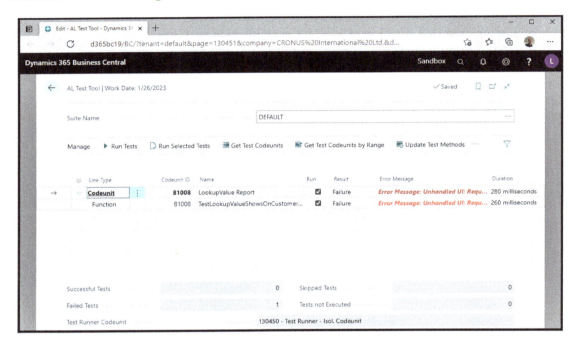

Figure 8.3 – Test example 8 having been run first time: RED

Ever thought you would be happy with **red**? But let's stay focused. This error doesn't mention anything about the missing column:

```
Unhandled UI: RequestPage 50000
```

Step 3 bis – Add just enough application code

You know, we could have foreseen it: the report opens with a request page, which is a UI element that indeed needs to be handled with the following UI handler (back to **Step 3**):

```
[RequestPageHandler]
procedure CustomerListRequestPageHandler(
  var CustomerListRequestPage: TestRequestPage CustomerList)
begin
  // Empty handler used to close the request page, default
```

```
    // settings are used
end;
```

And we have to make sure it can be used by our test, so we need to add the following attribute to our test function:

```
[HandlerFunctions('CustomerListRequestPageHandler')]
```

Step 4 bis – Run test seeing it probably fail

As we fixed the unhandled UI, let's build, deploy, and run the test again. **Step 4** once more: red!

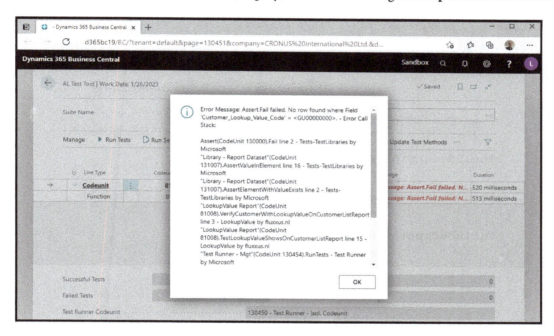

Figure 8.4 – Second run of test example 8 – RED

Hovering over the error message you can read the expected:

```
No row found where Field 'Customer_Lookup_Value_Code' =
<GU00000000>
```

Step 5 – Adjust the application code just enough to make it pass

Time to fulfill the customer wish: implement the Lookup Value Code column on the report as depicted in *Figure 8.2*.

A condensed version of the cloned report on this and the next page shows where we have added the Lookup Value Code field as a column to the report object:

```
report 50000 "CustomerList"
{
  . . .
  dataset
  {
    dataitem(Customer; Customer)
    {
      . . .
      column(Customer_Phone_No_; "Phone No.")
      {
        IncludeCaption = true;
      }
      column(Customer_Lookup_Value_Code;"Lookup Value Code")
      {
        IncludeCaption = true;
      }
      . . .
    }
  }
}
```

Build, deploy, and run: **greeeeeeeeen**!

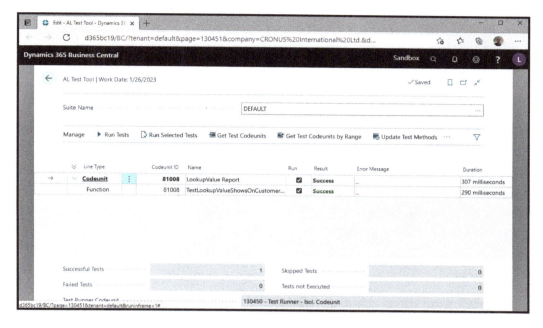

Figure 8.5 – Final run – before refactoring – of test example 8 – GREEN

This wasn't that hard, right? Using TDD for Business Central development, why not?

Step 6 – Refactor your code and rerun the test to prove all code is still well

Aha, **Step 6** of the *red-green-refactor* mantra: **Refactor your code and rerun the test to prove all code is still well**, isn't completed yet. As marked above `CreateCustomerWithLookupValue` and `CreateLookupValueCode` clearly carry a smell of code duplication.

CreateLookupValueCode

With the note *for implementation see Test example 1* in **Step 1**, I hinted for you to implement `CreateLookupValueCode` just as we did in *Test example 1*, not making use of the same method in the library yet:

```
local procedure CreateLookupValueCode(): Code[10]
// this smells like duplication ;-) - see test example 1
var
  LookupValue: Record LookupValue;
```

```
begin
  LookupValue.Init();
  LookupValue.Validate(
    Code,
    LibraryUtility.GenerateRandomCode(
      LookupValue.FieldNo(Code),
      Database::LookupValue));
  LookupValue.Validate(Description, LookupValue.Code);
  LookupValue.Insert();
  exit(LookupValue.Code);
end;
```

With what we did above under *Smell of duplication 1*, it's a piece of cake to fix this. As always, the final code is to be found on GitHub.

And as always too – rerun before the next code change: **green**. Notice that we now rerun all tests:

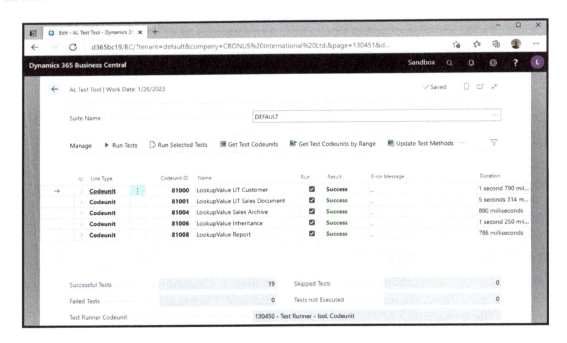

Figure 8.6 – Final run – after refactoring – of Test example 8 – GREEN

CreateCustomerWithLookupValue

With the note *have a look at scenario #0024 mentioned with Test example 6 this smells like duplication* in the code of `CreateCustomerWithLookupValue`, I pointed out that this helper function seems to resemble the helper function with the same name that we already created in codeunit `LookupValue Inheritance` (81006). This is its code:

```
local procedure CreateCustomerWithLookupValue(
  LookupValueCode: Code[10]): Code[20]
var
  Customer: Record Customer;
begin
  LibrarySales.CreateCustomer(Customer);
  Customer.Validate("Lookup Value Code", LookupValueCode);
  Customer.Modify();
  exit(Customer."No.");
end;
```

Where they both share the same method name and the first three lines of code, they deviate with respect to the signature. Trying to abstract both helper functions into one to eliminate the suspected duplication will only make things more complicated. It will not serve the goal of refactoring: getting *reusable*, *readable*, and *minimalistic* code. Let's dare to leave it as is.

> **Note**
>
> Implementing scenario #0029 did not necessitate a report layout to be defined as it only is about checking the dataset. To, however, be able to illustrate the purpose of having a dataset with *Figure 8.2* we surely needed a report layout – a major reason why you will find it in the GitHub repo.

Test the test

But before we move on, let's test our test!

Adjust the test so the verification errs

How about changing the `LookupValueCode` parameter when being called from the test function?

```
//[THEN] Report dataset contains both customers with lookup value
VerifyCustomerWithLookupValueOnCustomerListReport(
  Customer[1]."No.", 'LUC 1');
VerifyCustomerWithLookupValueOnCustomerListReport(
```

```
Customer[2]."No.", 'LUC 2');
```

This will throw the following error:

```
Assert.Fail failed. No row found where Field
   'Customer_Lookup_Value_Code' = <LUC 1>
```

As you might have discerned that only one error is thrown. This is the case with every test function: if an error is thrown the test will stop, and record the error unless **asserterror** is used. Thus, in the current case, the second call to VerifyCustomerWithLookupValueOnCustomerListReport will not occur. To test the second assert we will need to revert the first assert and rerun the test:

```
//[THEN] Report dataset contains both customers with lookup value
VerifyCustomerWithLookupValueOnCustomerListReport(
   Customer[1]."No.", Customer[1]."Lookup Value Code");
VerifyCustomerWithLookupValueOnCustomerListReport(
   Customer[2]."No.", 'LUC 2');
```

Now, the following error occurs:

```
Assert.Fail failed. No row found where Field
   'Customer_Lookup_Value_Code' = <LUC 2>
```

Test example 9 – how to test with permissions

When implementing the customer wish, the focus is on getting the resulting new functionality working, irrespective of if you go down the *test first* road or not. With testing, we verify if this new functionality is working as intended. This does, however, not necessarily mean that all Business Central users will be able to operate this new functionality as this also depends on the **permission sets** that have been assigned to them.

As you might know, the most basic level at which we define permissions is the **table data level**. When introducing a new table in your extension, like the Lookup Value table in the **LookupValue** extension, you also need to provide at least one permission set that defines the access to the table data. And like with any feature you provide; you should also test whether the permission set works right. How to do this that's what this test example is about.

In order to enable a user to access the data in the Lookup Value table, he or she needs to get this permission set assigned. More specifically to be able to **read (R)** records from, **insert (I)** records into, **modify (M)** records in, and **delete (D)** records from the Lookup Value table.

And, of course, if the user does not have this permission set assigned, he or she will not be able to access, that is, read, insert, modify, or delete, the data, and an error will be thrown. To confirm that this all works as it should we need to verify this with the following scenarios from our ATDD sheet:

```
[FEATURE] LookupValue Permissions
[SCENARIO #0041] Create lookup value without permissions
[SCENARIO #0042] Create lookup value with permissions
[SCENARIO #0043] Read lookup value without permissions
[SCENARIO #0044] Read lookup value with permissions
[SCENARIO #0045] Modify lookup value without permissions
[SCENARIO #0046] Modify lookup value with permissions
[SCENARIO #0047] Delete lookup value without permissions
[SCENARIO #0048] Delete lookup value with permissions
[SCENARIO #0049] Open Lookup Values Page without permissions
[SCENARIO #0050] Open Lookup Values Page with permissions
```

As we will later see, this set of tests is just a start.

For now, let's pick out one or two from this list and implement them the TDD way. The resulting code for the rest you can reference in the GitHub repo.

> **Note**
>
> The use of permission sets is a standard Business Central feature. If you need to learn more about it: https://docs.microsoft.com/en-us/dynamics365/business-central/ui-define-granular-permissions.

Step 1 – Take a test from the test list and write the test code

First things first, let's pick scenario #0041 from our test list:

```
[FEATURE] LookupValue Permissions
[SCENARIO #0041] Create lookup value without permissions
[GIVEN] Full base starting permissions
[WHEN] Create lookup value
[THEN] Insert permissions error thrown
```

Now how do we implement this?

Using Library - Lower Permissions

Experienced Business Central developers might know that the AL language and the Business Central application do not provide us with methods to set or manipulate the permissions the current user has been assigned. Therefore, before we can code away this test, we need to elaborate a little bit on the technical side of testing permissions.

If we want to be able to test a feature against a specific permissions setup for the user running the test, it should be possible to manipulate the user's permission during test code execution. Typically at the start of a test, when the test fixture will be set, the user should have the *Super* role assigned to allow for any data creation. Once the test gets to the exercise phase the permissions of the user should be restricted. A `.dll` called `PermissionTestHelper`, provided as a server add-in, was introduced to enable this manipulation of the user's permissions during (test) code execution. Note that, once the permissions have been restricted, `PermissionTestHelper` can only elevate permissions back to the level of the permissions of the user account that runs the tests. Thus, to cover the whole range of permissions, make sure to run your tests with a user account that has the *Super* role assigned.

> **Note**
>
> `PermissionTestHelper` will not change the permission setup of the user in the database. It will only change its permissions in the current session.
>
> As `PermissionTestHelper` is a .NET component you can only run these tests in an on-prem environment and not in SaaS. As a consequence, you need to set the `target` in your `app.json` to `OnPrem`.

For our tests, we do not need to reference `PermissionTestHelper` directly from our code. Instead, we will make use of the **wrapper** codeunit `Library - Lower Permissions` that resides in the `Tests-TestLibraries` app provided by Microsoft.

`Library - Lower Permissions` has four basic functions:

- `StartLoggingNAVPermissions([PermissionSetRoleID: Code[20]])`

  Instantiates the `PermissionTestHelper` and, optionally, sets the starting permission set to `PermissionSetRoleID`. Like the following example where the starting permission set will be *Super* for current user:

  ```
  LibraryLowerPermissions.StartLoggingNAVPermissions(
      'SUPER');
  ```

- `StopLoggingNAVPermissions()`

  Disposes of `PermissionTestHelper` instance and reverts the system to the previous state regarding permissions.

- `PushPermissionSet(PermissionSetRoleID: Code[20])`

  Clears all previous permission assignments and sets the permission set to `PermissionSetRoleID`. Like the following example where the permission set will be *D365 Bus Full Access* only for current user:

  ```
  LibraryLowerPermissions.PushPermissionSet(
      'D365 BUS FULL ACCESS');
  ```

- `AddPermissionSet(PermissionSetRoleID: Code[20])`

  Adds the permission set `PermissionSetRoleID` on top of the previously assigned permission sets. Like the following example where the permission *Lookup Value* will be added to all active permissions set for the current user:

  ```
  LibraryLowerPermissions.AddPermissionSet('LOOKUP VALUE');
  ```

Next to these, there is a whole collection of shortcut methods that pushes or adds a specific permission set, like:

- `SetO365Full()` that pushes the *D365 Bus Full Access* permission set.
- `SetCustomerEdit()` that pushes the *D365 Customer, Edit* permission set.
- `AddCustomerEdit()` that adds the *D365 Customer, View* permission set.

Now, let's apply our newly gained knowledge to implement scenario #0041. *Create, embed, write* will thus lead to the following:

```
codeunit 81020 "LookupValue Permissions"
{
  Subtype = Test;

  trigger OnRun()
  begin
    // [FEATURE] LookupValue Permissions
  end;

  [Test]
  procedure CreateLookupValueWithoutPermissions()
  begin
```

```
//[SCENARIO #0041] Create lookup value without permissions

//[GIVEN] Full base starting permissions
SetFullBaseStartingPermissions();

//[WHEN] Create lookup value
asserterror CreateLookupValueCode();

//[THEN] Insert permissions error thrown
VerifyInsertPermissionsErrorThrown();
    end;
}
```

I reckon by now CreateLookupValueCode doesn't need an explanation. It can be implemented exactly like we refactored it in *Test example 7*, reaping the benefit of this refactoring as we can now re-use the function instead of writing the code again

Let's therefore focus on the other helper functions SetFullBaseStartingPermissions and VerifyInsertPermissionsErrorThrown.

With SetFullBaseStartingPermissions we start the permissions logging and set the starting permissions set for our tests to *D365 Bus Full Access*:

```
local procedure SetFullBaseStartingPermissions()
begin
  LibraryLowerPermissions.StartLoggingNAVPermissions(
    'D365 BUS FULL ACCESS');
end;
```

We do this as *D365 Bus Full Access* is the standard permissions set that Microsoft uses when validating your extension for AppSource. As it is based on the standard application it does not include any permissions to the objects in the **LookupValue** extension. Running any of our tests that try to read, insert, modify, or delete data in the tables in our extension will fail due to failing permissions.

The expected failure that occurs in our test `CreateLookupValueWithoutPermissions` will be `'You do not have the following permissions on TableData Lookup Value: Insert'`. `VerifyInsertPermissionsErrorThrown` implements its:

```
local procedure VerifyInsertPermissionsErrorThrown()
var
   YouDoNotHavePermissions: Label 'You do not have the following
      permissions on TableData LookupValue: Insert';
begin
   Assert.ExpectedError(YouDoNotHavePermissions);
end;
```

> **Notes**
>
> (1) We use `asserterror` in the WHEN implementation as we are expecting an error to occur, when creating a lookup value code.
>
> (2) You might recall from *Test example 2* that a simple way to get the expected error message is to run the tests without the `asserterror`. This will make the test error out and throw the error message, that you can then easily copy and paste.

Step 2 – Compile test code yielding red as application code is not yet there, and Step 3 – Add just enough application code

Well, funny enough, nothing needs to be done here as the test code compiles, because the app code is already there. Indeed, not every phrase of the *red-green-refactor* mantra needs to be sung out loud.

Step 4 – Run the test seeing it probably fail

Once again the test code is ready, so, get it deployed and get the test code running: green!

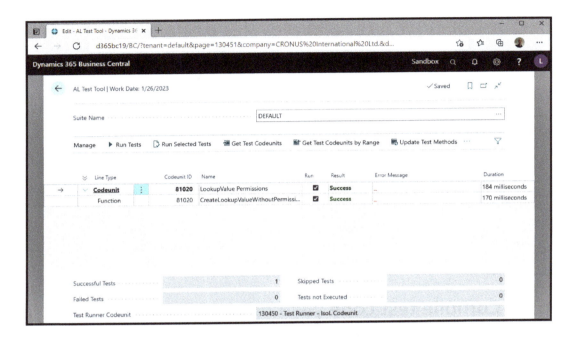

Figure 8.7 – Test example 9, the first scenario, having been run: GREEN

Who said we would see *it probably fail*?

Step 5 – Adjust application code just enough to make it pass, and Step 6 – Refactor your code and rerun the test to prove all code is still well

Do you see any arguments to adjust the application code, or refactor either app or test code? Me neither.

Step 7 – Move to the next test on the list and repeat from Step 1

We just picked the first test from the **LookupValue Permissions** feature test list, so, enough left to pick a next one.

Continue the red-green-refactor mantra

Indeed, there are more tests on our list and thus we can continue singing the *red-green-refactor* mantra by repeating from **Step 1** for the next test.

Step 1 – Take a test from the test list and write the test code

To balance the previous test, let's pick the next scenario from our test list: #0042.

```
[FEATURE] LookupValue Permissions
[SCENARIO #0042] Create lookup value with permissions
[GIVEN] Full base starting permissions extended with Lookup Value
[WHEN] Create lookup value
[THEN] Lookup value exists
```

Create, embed, write – will lead to the following:

```
[Test]
procedure CreateLookupValueWithPermissions()
var
    LookupValueCode: Code[10];
begin
    //[SCENARIO #0042] Create lookup value with permissions

    //[GIVEN] Full base permissions extended with Lookup Value
    SetFullBaseStartingPermissionsExtendedWithLookupValue();

    //[WHEN] Create lookup value
    LookupValueCode := CreateLookupValueCode();

    //[THEN] Lookup value exists
    VerifyLookupValueExists(LookupValueCode);
end;
```

Likewise, with `SetFullBaseStartingPermissionsExtended`
in scenario #0041 we start the permissions logging with
`SetFullBaseStartingPermissionsExtendedWithLookupValue` and set the starting
permissions set for our tests to *D365 Bus Full Access* and add the additional permissions with
respect to the `LookupValue` table:

```
local procedure
  SetFullBaseStartingPermissionsExtendedWithLookupValue()
begin
  LibraryLowerPermissions.StartLoggingNAVPermissions(
    'D365 BUS FULL ACCESS');
  LibraryLowerPermissions.AddPermissionSet('LOOKUP VALUE');
end;
```

`VerifyLookupValueExists` has to verify if the new lookup value record as created in the
WHEN clause indeed exists in the `LookupValue` table:

```
local procedure VerifyLookupValueExists(
  LookupValueCode: Code[10])
var
  LookupValue: Record LookupValue;
begin
  LookupValue.SetRange(Code, LookupValueCode);
  Assert.RecordIsNotEmpty(LookupValue);
end;
```

Step 2 – Compile test code yielding red as application code is not yet there, and Step 3 – Add just enough application code

Again, as with scenario #0041, nothing needs to be done here as the test code compiles, because
the app code is already there.

Step 4 – Run test seeing it probably fail

Deploy our code and run the tests. The first one, scenario #0041, of course succeeds, but our new one fails: **red**!

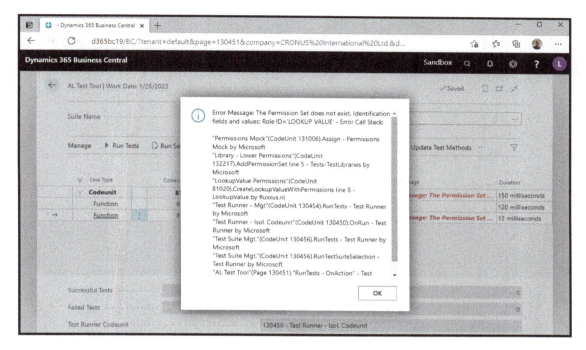

Figure 8.8 – Test example 9, the second scenario, having been run first time: RED

The error message clearly tells us what is missing:

```
The Permission Set does not exist. Identification fields and
values: Role ID='LOOKUP VALUE'
```

Step 5 – Adjust application code just enough to make it pass

Apparently, our extension does not contain a permission set *LOOKUP VALUE* and thus we need to provide it. This we do by defining the following permissionset object in our **LookupValue** extension:

```
permissionset 50000 "Lookup Value"
{
  Assignable = true;
  Caption = 'Lookup Value';
  Permissions =
```

```
    tabledata LookupValue = RIMD;
}
```

Deploy our code and run the tests: **green**!

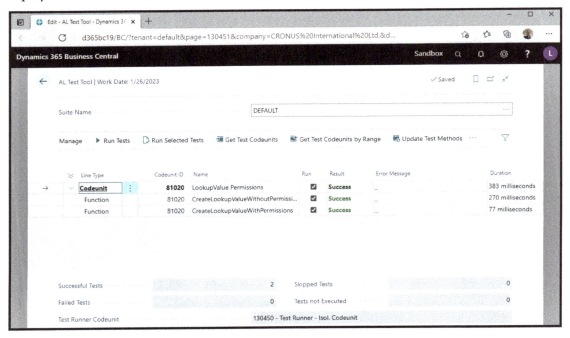

Figure 8.9 – Test example 9, the second scenario, having been run a second time: GREEN

Step 6 – Refactor your code and rerun the test to prove all code is still well

OK, do you see anything to refactor? Actually, I do, even though it might be easy to overlook or seem somewhat futile:

```
local procedure
  SetFullBaseStartingPermissionsExtendedWithLookupValue()
begin
  LibraryLowerPermissions.StartLoggingNAVPermissions(
    'D365 BUS FULL ACCESS');
  LibraryLowerPermissions.AddPermissionSet('LOOKUP VALUE');
end;
```

Do you notice?

Let's refactor this to the following:

```
local procedure
   SetFullBaseStartingPermissionsExtendedWithLookupValue()
begin
   SetFullBaseStartingPermissions();
   LibraryLowerPermissions.AddPermissionSet('LOOKUP VALUE');
end;
```

As the saying goes: "those who do not see the small things in life, can't really be happy." At the same time bear in mind that the TDD way is one of the small steps at a time and that clearly also goes for the refactoring part. Time to deploy and run, run, run, ruuuuuuun: **green**.

Step 7 – Move to the next test on the list and repeat from Step 1

Undeniably our test list is far from done. Creating, that is inserting, a new lookup value is just one of the data access modes that we need to verify. As the test list shows we also need to check the permissions for reading (scenarios #0043 and #0044), modifying (#0045 and #0046), and deleting (#0047 and #0048). I leave it to you to check the implementation of these six scenarios in the GitHub repo, including the two scenarios to check opening the Lookup Values page (#0049 and #0050).

But as noted above, there is even more to check with respect to permissions. Where the foregoing ten scenarios verify direct data access to the LookupValue table there is also indirect access to be checked. Any field, with a so-called **TableRelation**, placed on a page will only be rendered if the user has sufficient access permissions assigned. The consequence in the context of our LookupValue extensions is that we need to verify all our 36-page extensions that place the Lookup Value Code field on a standard page. Two tests for each page: *without* and *with* permissions. We'll only list the scenarios for the Customer Card and Customer List pages and refer to their implementation in the GitHub repo, also regarding the other 34 pages:

[SCENARIO #0051] Check lookup value on customer card without permissions

[SCENARIO #0052] Check lookup value on customer card with permission

[SCENARIO #0053] Check lookup value on customer list without permissions

[SCENARIO #0054] Check lookup value on customer list with permissions

Some notes on testing permissions and the version of Business Central

Testing permissions was first and foremost an internal Microsoft need. As such it has not been well documented and has not always worked as straightforwardly as you might expect. Microsoft even used a so-called internal test runner codeunit with their permissions tests. With the latest version of Business Central, that is Microsoft Dynamics 365 Business Central 2021 wave 2, aka BC19, they, however, have improved various parts with respect to permissions testing. While I was writing this book, Microsoft decided to downgrade `Permissions Mock` to Microsoft Dynamics 365 Business Central 2021 wave 1, aka BC18. It was released with BC18 cumulative update 3.

First of all, the use of the `PermissionTestHelper` component has been wrapped in a dedicated test library called `Permissions Mock`. Secondly, the logging of test permissions is initiated by default from any of the available test runners. The latter means that in our permissions tests we do not necessarily need to start the logging anymore, using the `StartLoggingNAVPermissions` method. It would therefore suffice to only push/set the base permissions set. I choose, nevertheless, to use the `StartLoggingNAVPermissions` method as this will also work on older versions of Business Central.

With the above, it might be clear that it's best to get into permission testing on BC18, cu3, and BC19.

Summary

In this chapter you built tests the TDD way, using the *red-green-refactor* mantra. But before you did that, you cleared the refactoring *backlog* for all the tests you have been constructing so far in the previous chapters. Based on this *clean slate* you applied TDD to testing a report dataset and permissions.

In the next chapter, *Chapter 9, How to Integrate Test Automation in Daily Development Practice*, we step into the fourth part of this book, in which we will discuss how to integrate your test automation in your day-to-day development practice, including the tests provided by Microsoft.

Section 4: Integrating Automated Tests in Your Daily Development Practice

This section elaborates on a number of best practices that might turn out to be beneficial for you and your team in getting test automation working in your day-to-day work, and how to extend your test practices by making use of the standard test collateral Microsoft provides with Dynamics 365 Business Central.

This section contains the following chapters:

9

How to Integrate Test Automation in Daily Development Practice

You've got this far into this book, so by now you have a clear notion of the needs for, and benefits of, test automation for Dynamics 365 Business Central. You have also already exercised designing and writing tests based on *Section 3, Designing and Building Automated Tests for Microsoft Dynamics 365 Business Central*. The next step is to bring into practice what you have learned.

Reading the book so far, trying to understand the matters discussed, and working on your first exercises, it might still be a threshold to take to make this all part of your daily work. As mentioned earlier, test automation is a team effort. Therefore, in this chapter, we will elaborate on a number of best practices that might turn out to be beneficial for you and your team in getting test automation working. As such, we will cover the following topics:

- Casting the customer wish into ATDD scenarios
- Learning and improving by taking small steps
- Making the test tool, and some more, your friend
- Maintaining your test code
- Organizing your extensions
- Integrating with the daily build system

Technical requirements

Like in the previous chapter, we will refer to the **LookupValue** extension. You can find its code on GitHub: `https://github.com/PacktPublishing/Automated-Testing-in-Microsoft-Dynamics-365-Business-Central-Second-Edition`.

In this chapter, we will continue with the code from *Chapter 8, From Customer Wish to Test Automation – the TDD way*: `https://github.com/PacktPublishing/Automated-Testing-in-Microsoft-Dynamics-365-Business-Central-Second-Edition/tree/main/Chapter 08 (LookupValue Extension)`.

The final code for this chapter can be found in `https://github.com/PacktPublishing/Automated-Testing-in-Microsoft-Dynamics-365-Business-Central-Second-Edition/tree/main/Chapter 09 (LookupValue Extension)`.

Details on how to use this repository and how to set up VS Code are discussed in *Appendix, Getting Up and Running with Business Central, VS Code, and the GitHub Project*.

Casting the customer wish into ATDD scenarios

Crucial to getting test automation into your daily development practices is the adoption of it by the team. Like requirements and application code, tests and test code should be owned by the development team; not just formally, but also actively. Good application code does not emerge from a single-lined customer wish; it derives from an eventually well-detailed and formalized customer wish, aka requirements. And the same applies to tests and test code.

As discussed in *Chapter 5, Test Plan and Test Design*, formalize your requirements by using the ATDD design pattern. Cast your customer wish in ATDD scenarios. Break down each wish into a list of tests and make this your primary vehicle of communication for (1) detailing of your **customer wish**, (2) implementation of your **application code**, (3) structured execution of your **manual tests**, (4) coding of your **test automation**, and (5) up-to-date **documentation** of your solution. Your test automation will be a logical result of all previous work.

As developers are the ones that will be doing both the application and test coding, and in general are not the ones that understand the customer wish the best, ATDD scenarios already hit *two out of five birds with one stone*. Make use of my ATDD Scenarios Excel sheet on GitHub, called `Clean ATDD sheet.xlsx`, in the folder `Excel sheets\Examples` to get your team to start casting customer wishes into ATDD scenarios. It's exactly what I did when working on the test examples from *Chapter 6* through *Chapter 8*. See the following screenshot for an impression:

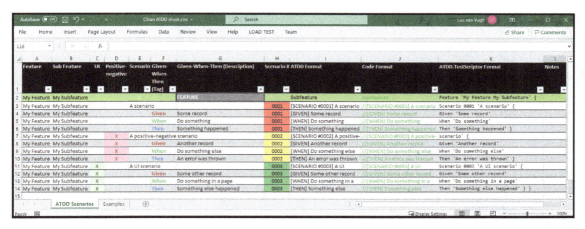

Figure 9.1 – The Clean ATDD sheet

The first eight columns are for the product owners, functional consultants, and/or key users to fill in:

- `Feature`
- `Sub Feature`
- `UI`
- `Positive-negative`
- `Scenario`
- `Given-When-Then (Tag)`
- `Given-When-Then (Description)`
- `Scenario #`

The next three columns, `ATDD Format`, `Code Format`, and `ATDD.TestScriptor Format`, are automatically populated, as also shown in *Figure 9.1*. Using the **4-steps recipe** to *create*, *embed*, *write*, and *construct* the test code, it's the `Code Format` column you will be gladly making use of as it contains *the green*, that is, the formatted and already out-commented GIVEN-WHEN-THEN scenarios. These are ready to be copied and pasted into your test codeunit and embedded into your test functions.

ATDD.TestScriptor

As the *4-steps recipe* is – right – a *recipe*, it's an obvious candidate for automation. Given ATDD scenarios, the subsequent steps defined by the *4-steps recipe* are always the same:

1. Create a test codeunit, with a name based on the FEATURE tag.
2. Embed the customer wish into a test function with a name based on the SCENARIO tag.

3.   Write your test story, based upon **GIVEN**, **WHEN**, and **THEN** tags.

4.   Construct your real code.

It's only the fourth and final step that needs the real intelligence of the developer.

PowerShell

When working on the first edition of this book, the idea arose of automating the first three steps of the *4-steps recipe*. Jan Hoek helped me build a **PowerShell** module that became known as the **ATDD.TestScriptor** and can be installed from the PowerShell Gallery: `http://powershellgallery.com/packages/ATDD.TestScriptor`. To allow you to run PowerShell scripts in VS Code – which we will do below – make sure that you have added an extension for that to VS Code, like `https://marketplace.visualstudio.com/items?itemName=ms-vscode.PowerShell`.

This ATDD.TestScriptor PowerShell module defines cmdlets for the five ATDD tags and, with these, lets you write ATDD scenarios. The exact format is the one you find in the `ATDD.TestScriptor Format` column of the ATDD Excel sheet (see *Figure 9.2*):

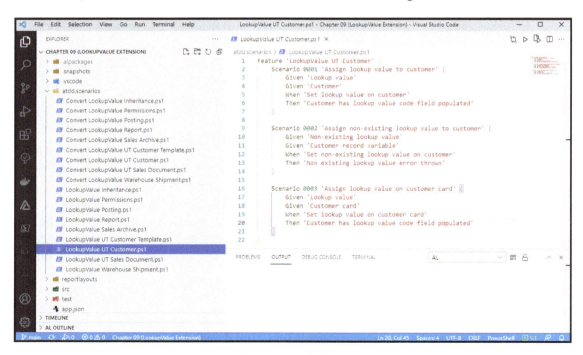

Figure 9.2 – An atdd.scenarios folder in VS Code project

The basic process to get from your ATDD scenarios to AL code using the **ATDD.TestScriptor** is as follows:

1. Copy, paste, and save the content of the `ATDD.TestScriptor Format` column into a PowerShell file.

2. Use the `ConvertTo-ALTestCodeunit` cmdlet of the module to convert this into a `.al` test code structure.

For each of these two steps, I create a PowerShell file which I store in an `atdd.scenarios` folder in my VS Code project folder:

1. `<ATDD feature>.ps1`

2. `Convert <ATDD feature>.ps1`

> **Note**
>
> Before you are going to use the **ATDD.TestScriptor** in the following example, note that this module will only work right if the extension folder is in the root of your VS Code workspace. It will, for example, not work if you use a **multi-root workspace**, discussed below in *Organizing your extensions*.

Let's have a look at how this works using the *LookupValue Posting* feature for which we have not created test code yet:

```
[FEATURE] LookupValue Posting
[SCENARIO #0022] Posted sales invoice and shipment inherit lookup
                 value from sales order
[GIVEN] Sales order with lookup value
[WHEN] Post sales order (invoice & ship)
[THEN] Posted sales invoice has lookup value from sales order
[THEN] Sales shipment has lookup value from sales order
[SCENARIO #0027] Posting throws error on sales order with empty
                 lookup value
[GIVEN] Sales order without lookup value
[WHEN] Post sales order (invoice & ship)
[THEN] Missing lookup value on sales order error thrown
[SCENARIO #0023] Posted warehouse shipment line inherits lookup
                 value from sales order
[GIVEN] Location with require shipment
[GIVEN] Warehouse employee for current user
```

[GIVEN] Warehouse shipment line from sales order with lookup
        value
[WHEN] Post warehouse shipment
[THEN] Posted warehouse shipment line has lookup value from
       sales order
[SCENARIO #0025] Posting throws error on warehouse shipment line
                 with empty lookup value
[GIVEN] Location with require shipment
[GIVEN] Warehouse employee for current user
[GIVEN] Warehouse shipment line from sales order without lookup
        value
[WHEN] Post warehouse shipment
[THEN] Missing lookup value on sales order error thrown

> **Note**
>
> You find these ATDD descriptions for the *Lookup Value Posting* feature in the ATDD
> sheet on the GitHub repo: https://github.com/PacktPublishing/
> Automated-Testing-in-Microsoft-Dynamics-365-Business-
> Central-Second-Edition/blob/main/Excel sheets/ATDD
> Scenarios/LookupValue.xlsx.

This is what you have to do step by step:

1. In the atdd.scenarios folder of the VS Code project, create a file to hold the ATDD
    scenario(s). In this context, the file will be called LookupValue Posting.ps1.

    If this folder does not exist yet, create it.

2. In the ATDD sheet, copy from the ATDD.TestScriptor Format column the lines
    that belong to the *Lookup Value Posting* feature, including the **FEATURE** line.

3. Paste these lines into the .ps1 file created in *step 1*:

    ```
    [Feature] 'LookupValue Posting' {
    [Scenario] 0022 'Posted sales invoice and shipment inherit
    lookup value from sales order' {
    [Given] 'Sales order with lookup value'
    [When] 'Post sales order (invoice & ship)'
    [Then] 'Posted sales invoice has lookup value from sales
    order'
    [Then] 'Sales shipment has lookup value from sales order' }
    ```

```
[Scenario] 0027 'Posting throws error on sales order with
empty lookup value' {
[Given] 'Sales order without lookup value'
[When] 'Post sales order (invoice & ship)'
[Then] 'Missing lookup value on sales order error thrown' }
[Scenario] 0023 'Posted warehouse shipment line inherits
lookup value from sales order' {
[Given] 'Location with require shipment'
[Given] 'Warehouse employee for current user'
[Given] 'Warehouse shipment line from sales order with lookup
value'
[When] 'Post warehouse shipment'
[Then] 'Posted warehouse shipment line has lookup value from
sales order' }
[Scenario] 0025 'Posting throws error on warehouse shipment
line with empty lookup value {
[Given] 'Location with require shipment'
[Given] 'Warehouse employee for current user'
[Given] 'Warehouse shipment line from sales order without
lookup value'
[When] 'Post warehouse shipment'
[Then] 'Missing lookup value on sales order error thrown' } }
```

4.   You might want to format the text so it displays better, using the *Format Document* feature of VS Code first, putting each closing curly bracket on the next line, and maybe wrap the description that is too long to the next line:

```
[Feature] 'LookupValue Posting' {
  [Scenario] 0022 'Posted sales invoice and shipment inherit
                   lookup value from sales order' {
    [Given] 'Sales order with lookup value'
    [When] 'Post sales order (invoice & ship)'
    [Then] 'Posted sales invoice has lookup value from sales
            order'
    [Then] 'Sales shipment has lookup value from sales order'
  }
  [Scenario 0027] 'Posting throws error on sales order with
                   empty lookup value' {
    [Given] 'Sales order without lookup value'
```

```
    [When] 'Post sales order (invoice & ship)'
    [Then] 'Missing lookup value on sales order error thrown'
}
[Scenario 0023] 'Posted warehouse shipment line inherits
                lookup value from sales order through
                warehouse shipment line' {
    [Given] 'Location with require shipment'
    [Given] 'Warehouse employee for current user'
    [Given] 'Warehouse shipment line from sales order with
            lookup value'
    [When] 'Post warehouse shipment'
    [Then] 'Posted warehouse shipment line has lookup value
            from warehouse shipment line'
}
[Scenario 0025] 'Posting throws error on warehouse shipment
                line with empty lookup value' {
    [Given] 'Location with require shipment'
    [Given] 'Warehouse employee for current user'
    [Given] 'Warehouse shipment line from sales order
            without lookup value'
    [When] 'Post warehouse shipment'
    [Then] 'Missing lookup value on sales order error thrown'
}
}
```

Find my resulting file here: https://github.com/PacktPublishing/
Automated-Testing-in-Microsoft-Dynamics-365-Business-Central-
Second-Edition/blob/main/Chapter 09 (LookupValue Extension)/
atdd.scenarios/LookupValue Posting.ps1.

5. Now create the second .ps1 file in the same directory, called Convert LookupValue Posting.ps1. In this file, we add the next PowerShell script.

6. The content of Convert LookupValue Posting.ps1 could be as follows:

```
& "./atdd.scenarios/LookupValue Posting.ps1" | '
  ConvertTo-ALTestCodeunit '
    -CodeunitID 81005 '
    -CodeunitName 'LookupValue Posting' '
    -InitializeFunction '
```

```
      -GivenFunctionName "Create {0}" '
      -ThenFunctionName "Verify {0}" '
      | Out-File '.\test\LookupValuePosting_2.Codeunit.al'
```

This script picks up the ATDD feature from the .ps1 file you just created in *step 1* and feeds that to the ConvertTo-ALTestCodeunit cmdlet. The ConvertTo-ALTestCodeunit cmdlet will then convert the ATDD scenarios into a .al test codeunit containing a test function for each ATDD scenario as per the first three steps of the *4-steps recipe*.

Find my *conversion* file here: https://github.com/PacktPublishing/Automated-Testing-in-Microsoft-Dynamics-365-Business-Central-Second-Edition/blob/main/Chapter 09 (LookupValue Extension)/atdd.scenarios/Convert LookupValue Posting.ps1.

7. Running this script will create the file LookupValuePosting_2.Codeunit.al, of which I stored a version on GitHub: https://github.com/PacktPublishing/Automated-Testing-in-Microsoft-Dynamics-365-Business-Central-Second-Edition/blob/main/Chapter 09 (LookupValue Extension)/test/LookupValuePosting_2.Codeunit.

Note that the .al file extension has been removed, so, the codeunit it contains is not picked up by the AL compiler.

A short description of the content of LookupValuePosting_2.Codeunit

LookupValuePosting_2.Codeunit contains:

- A test codeunit with ID 81005 and name LookupValue Posting.

  ID and name are determined by the CodeunitID and CodeunitName parameters of ConvertTo-ALTestCodeunit.

- Four test functions determined by the four scenarios in the ATDD feature file, LookupValue Posting.ps1.

- An Initialize method determined by the InitializeFunction parameter of ConvertTo-ALTestCodeunit. A call to it is added to each of the four test functions.

- The two remaining parameters of ConvertTo-ALTestCodeunit, GivenFunctionName, and ThenFunctionName, determine that each GIVEN and THEN tag will result in a function (call) whose name starts with Create or Verify, respectively.

> **Note**
>
> A description of the full set of parameters of the `ConvertTo-ALTestCodeunit` cmdlet can be accessed through `Get-Help`: `https://docs.microsoft.com/en-us/powershell/module/microsoft.powershell.core/get-help?view=powershell-7.1#description`. (and also in its GitHub repo: `https://github.com/fluxxus-nl/ATDD.TestScriptor/blob/master/docs/ConvertTo-ALTestCodeunit.md`).

Given `LookupValuePosting_2.Codeunit`, I wrote the *real code* in each of the pre-created helper methods, thus completing *step 4* of the *4-steps recipe*. The final result is the `.al` object file, `LookupValuePosting.Codeunit.al`: `https://github.com/PacktPublishing/Automated-Testing-in-Microsoft-Dynamics-365-Business-Central-Second-Edition/blob/main/Chapter 09 (LookupValue Extension)/test/LookupValuePosting.Codeunit.al`. Go there and study its content. You might even want to download both `LookupValuePosting_2.Codeunit` and `LookupValuePosting.Codeunit.al` and set them side by side with a text file compare tool to study the additions and changes I made to the final test codeunit. Note that each helper function created by the `ATDD.TestScriptor` cmdlet, `ConvertTo-ALTestCodeunit`, contains one default error statement, like, for example, `CreateLocationWithRequireShipment`:

```
local procedure CreateLocationWithRequireShipment()
begin
  Error('CreateLocationWithRequireShipment not implemented.')
end;
```

In the remaining part of this book, we will make use of the `ATDD.TestScriptor` PowerShell module to convert the ATDD scenarios into `.al` test codeunits.

> **Note**
>
> If you need a more comprehensive demo of the `ATDD.TestScriptor` PowerShell module, have a look at this Areopa webinar: `https://www.youtube.com/watch?v=ma48oWYWCvw`.

VS Code

Where the **ATDD.TestScriptor** PowerShell module can be a very effective tool for converting ATDD scenarios into a first version of `.al` test code, it's not a lot of help in maintaining existing `.al` test code. Therefore, I started a new project called **ATDD.TestScriptor for AL**, with the indispensable help of Márton Sági and David Feldhoff, building a **VS Code** extension that allows you to:

- abstract ATDD definitions from AL test codeunits
- create ATDD definitions in AL test codeunits
- remove ATDD definitions from AL test codeunits
- update ATDD definitions in AL test codeunits

Eventually, although not yet implemented, you will be able to import ATDD scenarios from an Excel sheet, and also abstract them from AL code into an Excel sheet. This way, the single source of truth will always be the AL code:

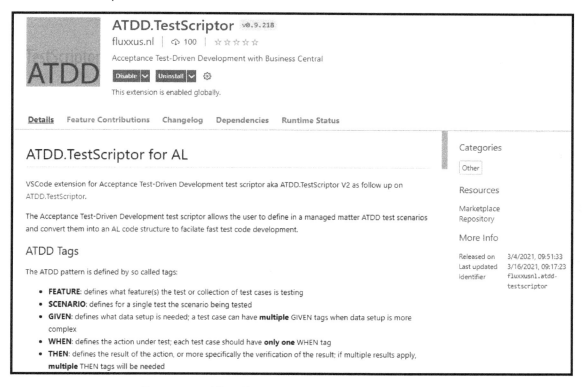

Figure 9.3 – ATDD.TestScriptor extension page in VS Code

The `ATDD.TestScriptor for AL` VS Code extension can be downloaded from `https://marketplace.visualstudio.com/items?itemName=fluxxusnl.atdd-testscriptor`. Here, you also can find a comprehensive description of the extension.

> **Note**
>
> A short demo of the `ATDD.TestScriptor for AL` VS Code extension was given as part of an Areopa webinar. You can find the recording of the demo here: `https://www.youtube.com/watch?v=om_LzWF-uyc`.

Opening the ATDD.TestScriptor in VS Code, by typing `ATDD.TestScriptor` in the command palette in our **LookupValue** project, will abstract all the tests present into a list in the **ATDD.TestScriptor** pane, as displayed in *Figure 9.4*:

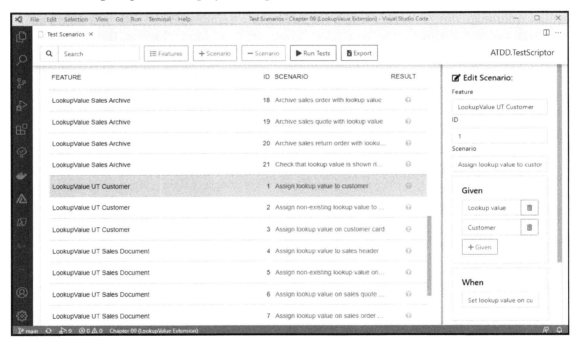

Figure 9.4 – ATDD.TestScriptor for AL

Learning and improving by taking small steps

You get to master test automation one step at a time. Learn and improve by doing the following:

- Choose one of the when-to-use-automated-testing options as discussed in *Chapter 1, Introduction to Automated Testing*, that fits you and your team best, and/or gives the best return on investment, to start mastering test automation.

- Start casting the customer wish into the scenarios that come to mind. Try to keep it simple. Preferably, you would like to get full coverage right away, but as it is a team effort, chances are high that holes will be identified before implementation starts. And if they will not be seen, chances are even bigger that you will hit upon these holes when starting to implement the code.

- Make use of my *4-steps recipe* (*create*, *embed*, *write*, and *construct*) for the conception of the test code. Let the ATDD.TestScriptor module and/or extension make this work even more efficient.

- Get the test(s) run with every step performed, and as soon as the code is deployable. Do not wait until you're finished but verify your effort immediately, at every step you take. See how your tests move from red to green.

- Take joy in *testing the test* as a last small step in completing it. Either verify the data created or, often much easier, adjust the test so the verification errs.

- Implement one scenario after the other and find yourself duplicating code parts. Do not force yourself into directly abstracting them into helper methods in libraries. Once a new test runs successfully, be in **TDD** mode, if you weren't already, and refactor to eliminate the duplication (the second rule of TDD). Remember that with TDD, we do not design *reusable*, *readable*, and *minimalistic* code upfront. Instead, rule 2 of **TDD** leads you there.

- Run the tests regularly once the application code and tests are in and when the next update to the feature is made. While implementing, do not wait until the code is ready, verify each atomic change by running the tests, and add new tests for the new and updated scenarios.

As the saying goes, *Rome wasn't built in a day*. Likewise, learning to master test automation isn't an instant achievement. It needs time and endeavor.

Making the test tool, and some more, your friend

In *Chapter 4, The Test Tools, Standard Tests, and Standard Test Libraries*, we introduced you to the test tool and we used it frequently during the work we did on the test examples. We applied it to *test the test* after having inserted a bug to deliberately cause a *verification error*. Next to VS Code, your coding tool, and debugger, the **test tool** is one of your best friends. Keep it running while you develop, and *do not wait till the code is ready*; *verify each atomic change by running the tests*, as mentioned previously. The goal is to make automated testing part of development.

Create a specific **test suite** to hold the test codeunits that relate to the code you are working on. In most of your projects, it's very likely, as with the *LookupValue* extension, that you end up with a bunch of test codeunits that will be executed within less than a minute. While working on a new test codeunit, create a new test suite to only hold that codeunit, and repeatedly carry it out until coding is done.

In this section, I would like to share a few more tools that come in handy in your daily work on tests.

Extending the test tool

When I started using the test tool more intensively years ago, there was one major omission for which I decided to build a simple extension. Once you have set up a test suite and run all the tests, you might end up having only a portion of the tests that passed successfully, and, logically, the other part not passing.

While finding and fixing the cause of the failures, you presumably would only want to run the faulty ones. The standard test tool lets you only activate/deactivate tests by manually checking/unchecking the Run field on the relevant function lines. Checking/unchecking Run on a codeunit line also does the same on all the function lines it entails. Manually setting this property in a large test suite can be very time-consuming. The standard test tool does not have the ability to quickly select everything based on test results, so I created an extension that adds those capabilities.

The test tool extension empowers you to *check* the Run field on the following:

- All tests
- Only failing tests, and thus disabling all others
- On non-failing tests, thus disabling failing tests

And, as a fourth option, to *deactivate* all tests, all within the filters applied to the test suite:

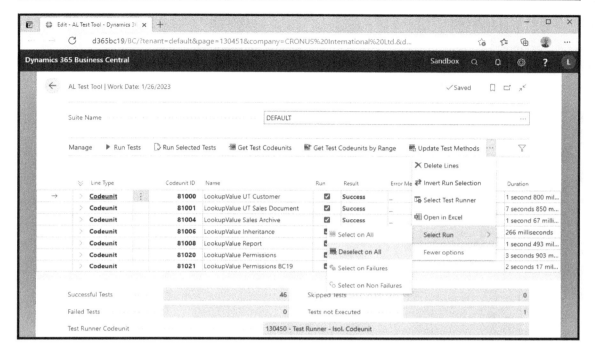

Figure 9.5 – The four actions of the Test Tool extension

For those four checks, the following four actions have been added as shown in *Figure 9.5*:

- **Select on All**
- **Deselect on All**
- **Select on Failures**
- **Select on Non Failures**

Hence, when using, for example, the **Select on Failures** action, all failing tests will have Run checked, while all other tests will have it unchecked.

The source of the test tool extension can be downloaded from this GitHub repository: `https://github.com/fluxxus-nl/Test-Tool-Extension`.

> **Note**
> Did you notice the AL test tool action, **Invert Run Selection**? It does what it says: it will invert the value of the **Run** field on the lines that are selected in the test tool.
> In my humble opinion, it is only useful when a small number of tests reside in your test suite.

AL Test Runner

Even though the test tool will be one of your best friends, it might not always be as efficient as you would like it to be. Working on a test in VS Code, you may want to instantly run the test, or multiple tests, without needing to shift to the Business Central web client:

Figure 9.6 – Al Test Runner extension page in VS Code

In this case, a very nice VS Code AL extension can help you out. With **AL Test Runner**, as it is called, you can run the test straight from VS Code with the results being displayed in VS Code. No need to switch to the web client as everything is handled from and in VS Code. There are, however, two major requirements for using this VS Code extension:

- You can only use it when you are developing against a Business Central Docker installation.

- You need to have the PowerShell module **BcContainerHelper** installed.

> **Note**
>
> If you do not use Docker yet, this is another reason to start. In my humble opinion, it has so many advantages to do so anyway. In all honesty, it took me a while before I passed the threshold of working with Docker two years ago, but it was more than worth doing that.

Using AL Test Runner

As the AL Test Runner extension includes a comprehensive description of its different features, I have no intention to show them all. Only to draw your attention a little bit more, this is how you basically can use the AL Test Runner.

If you have the AL Test Runner VS Code extension installed and you open a test codeunit object file, you might notice the following two things which are also displayed in *Figure 9.7*:

- Each test function name is highlighted with a color. If a test has not been executed yet with the AL Test Runner, it will be *yellowish*.

- Each test function will get two hyperlink actions assigned: **Run Test** and **Debug Test**. With these, you can run the test now or get it run while a debug session is simultaneously started:

Figure 9.7 – AL Test Runner actions and coloring shown in a test codeunit

To run a test with AL Test Runner:

1. Open the test codeunit object file that holds the test in VS Code.

2. Click on the **Run Test** hyperlink above the name of the test function. Let's do it for one of the tests shown in *Figure 9.7*: `ArchiveSalesOrderWithLookupValue`.

   When running for the first time, you will probably get the message **Please run the command again to run the test(s)**.

   Once AL Test Runner is running, you can follow its execution in the **terminal** pane in VS Code.

When ready, the executed test function either is highlighted red (failure) or green (success) and a resume of the test run will be displayed in the **output** pane. See also *Figure 9.8*:

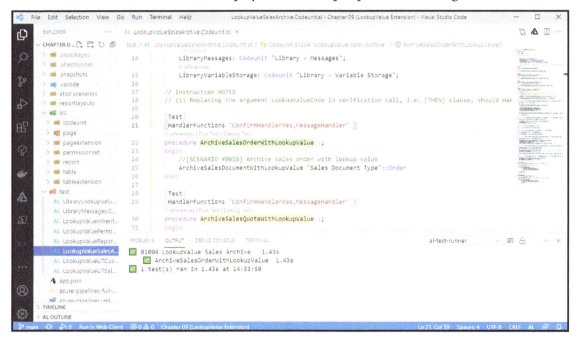

Figure 9.8 – Results of running one test with AL Test Runner; notice the changed coloring and the resume

Of course, before you can run a test, the test project needs to be deployed. Make sure to redeploy in case the code was updated after the last deployment.

Running a test, or multiple tests in the same test codeunit or in all other test codeunits can also be triggered from the VS Code command palette using:

- `AL Test Runner: Run Current Test`
- `AL Test Runner: Run All Tests`

- `AL Test Runner: Run Tests in Current Codeunit`

Notes

(1) AL Test Runner can be installed in VS Code from the Visual Studio marketplace: `https://marketplace.visualstudio.com/items?itemName=jamespearson.al-test-runner`.

(2) Download BcContainerHelper from `https://www.powershellgallery.com/packages/BcContainerHelper`.

Code Coverage

With code coverage, we want to express how much of our code is hit when executing a code, irrespective of it being a regular process or test. To be able to measure *how much of our code is hit*, a so-called code coverage tool is needed that records the code being executed and calculates the hit percentage of the total code in the `.al` objects being monitored. The Business Central base application provides us with one that we can open using the **Tell me what you want to do** feature. *Figure 9.9* shows the initial state of the Business Central **Code Coverage** tool not yet displaying any data:

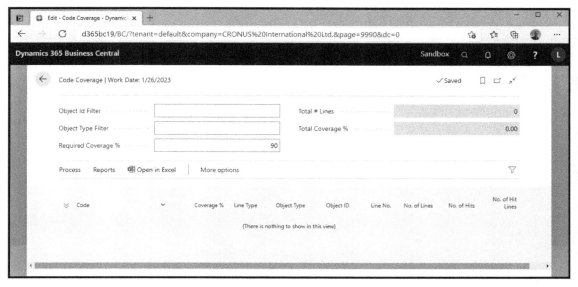

Figure 9.9 – Code Coverage tool in the initial state

To record the code you want to execute, you have to open a second web browser tab and open the page from where you will trigger this code. Before this is code is triggered, you need to start your code coverage tool. Once it has been carried out, you should stop the tool.

Running the Code Coverage tool

Let's do a recording of our tests and have a look at the result:

1. Open Business Central web client.

2. Open the **Code Coverage** page using the **Tell me what you want to do** feature.

3. Open another Business Central station in a second web browser tab.

4. Open the **AL Test Tool** page using the **Tell me what you want to do** feature.

5. Return to the **Code Coverage** page in the first web browser tab.

6. Start the tool by selecting **Process | Start**.

7. Return to the **AL Test Tool** page in the second web browser tab.

8. Run all tests by selecting **Run Tests** and click **OK** in the dialog.

9. Once the test run is completed, return to the **Code Coverage** page in the first web browser tab.

10. Stop the tool by selecting **Process | Stop**.

The lines in the tool will now display all objects that were touched during the test run and all their code lines. You will notice that this is quite a long list, probably much longer than you might have expected.

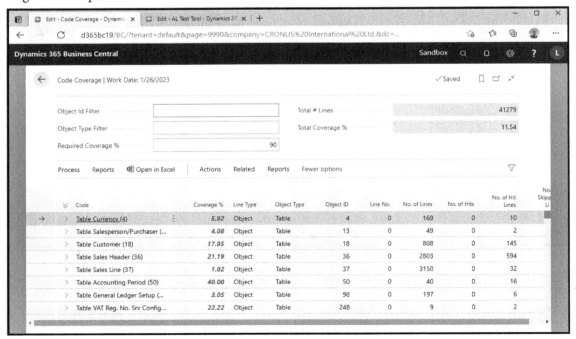

Figure 9.10 – Code Coverage tool with the recorded data after the test run

In the specific case of running all our tests that have been built so far for the *LookupValue* extension, *235* objects are involved with over *40,000* lines of code; see *Figure 9.10*. A major part of these objects being standard Business Central objects – only a very small number that belongs to our extension. The tool calculates that approx. 11.5 % of those lines have been hit by our tests. Even though we might be impressed by these numbers, we need to find out what they tell us.

Analyzing the code coverage result

When analyzing the result of a code coverage recording, it is important to have the right **focus**, and also realize the **constraints** of the recording.

Focus

When measuring the code being covered, your primary interest will be the code you are responsible for. In the context of executing automated tests against our *LookupValue* extension, we should only be interested in the code that is part of this extension. As such, we should apply a filter on the **Code Coverage** page:

- Set the **Object Id Type** field to `50000..81099` as this is the range of our extension.

    Result: 12 objects, *300* **Total Lines**, and a *99.67* **Total Coverage %** (see *Figure 9.11*):

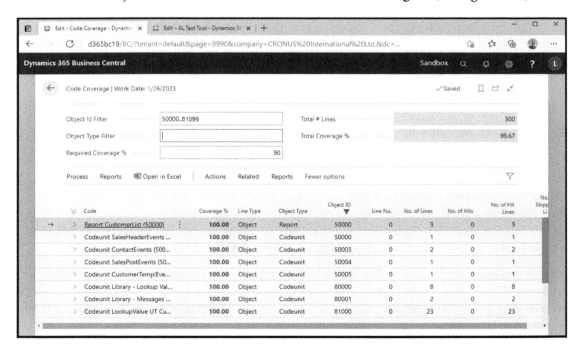

Figure 9.11 – Code Coverage tool with the recorded extension objects

As can be seen in *Figure 9.11*, the subset of objects includes both applications and test objects in our extension. Measuring the coverage of the test code, however, is of no direct interest. Therefore, we need to narrow our focus a little bit more.

- Set the **Object Id Type** field to `50000..50099` as this is the range that holds the application objects of our extension.

    Result: *5* objects, *9* lines of code, and *88.89%* coverage. Notice that all the lines display a coverage percentage of *100*. A quick analysis of the number shows that a total of *8* lines have been hit (see the **No. of Hit Lines** column on the lines) on a total of *9* lines (see **Total # Lines** on the header).

Inspecting the lines in the tool, we can conclude that all have been hit. At the same time, you might notice that it does not cover that a field has been accessed – more specifically, the `Lookup Value Code` field in the various page and table extensions.

Constraints

There are a number of constraints to the meaning of a code coverage recording and its coverage percentage:

- The analysis above reveals that the **Code Coverage** tool only records the code being hit, that is, the AL code to be found in triggers and functions. It does not record the usage of other parts of the AL objects, such as table fields and page controls.

- The recording of the code being hit only tells us what code was hit but does not tell us in what order. This could mean that, even though all code was hit, not all functional scenarios were covered. It also does not discern between often used scenarios and the rare ones.

- In this light, making your goal a *100%* coverage isn't a very meaningful, not to say a useless and costly endeavor. *80%* coverage is often reckoned to be adequate and achievable within a reasonable timeframe and at a realistic effort.

All in all, code coverage isn't an unambiguous measure. It has to be used thoughtfully in conjunction with a good understanding of the functionality and, based on the latter, a sufficient set of valid automated tests. Some even say that code coverage as such has no value at all except for the fact that, with a certain coverage, there's still a part of the code not covered.

This raises the question as to why this code wasn't covered and whether we should make the effort to get it covered. For that matter, when working on test automation for existing application code, measuring the code coverage will provide you with an insight into how your test coverage is progressing. It might be helpful to set a code coverage bar for every next deadline, like *35%* coverage for the first deadline, *50%* for the next, and so on.

> **Note**
> The AL Test Runner also includes a way of collecting code coverage measurements. Read more about it in this blog post: `https://jpearson.blog/2021/02/07/measuring-code-coverage-in-business-central-with-al-test-runner`.

Maintaining your test code

Like application code, test code is also code, so handle it as application code should be handled. This means:

- Test code should be secured by means of a **source code management** tool, providing you a history of changes and an easy way to revert to any previous state.

- Test code will most likely need to be updated with any new customer wish, as this will result in a change of the behavior of your solution.

- Test code most probably needs to be debugged, whether you like it or not, as any coding done by a developer potentially inserts new bugs, in both the application and the test code.

- Test code should be reviewed to ensure that, like application code, it meets the coding standards.

- And of course, test code should be stored with the application code to assure they're closely tied and in sync.

Next to this, when allowing other parties to extend your extension, empower them by providing access to your test code, like Microsoft sharing their tests with us.

Organizing your extensions

Starting with a new project, like we did with the *LookupValue* extension, I tend to have both the application and the test code in one VS Code folder just for one reason: *practicality*. This way, I only have to deploy one extension when verifying my work after an addition or change in either application or test code – or both, although TDD teaches us to do one or the other. This helps me to get my work done more efficiently.

Once the code is ready, however, we need to make sure that the application and test code end up in separate extensions – but still in the same repo – with the logical consequence that the test app will have a dependency on the application app. This way, we prevent tests from ending up, and possibly getting run, in a production environment. For this very reason, standard CRONUS does not contain the standard test codeunits – and test helper libraries.

Sure, if test codeunits are run within the test isolation of a test runner, no changes to the data will sustain. For that matter, the worst that could happen is that users will bump into locks. But what if a test codeunit accidentally gets run outside the isolation of a test runner, and test data is committed to the production system? Your client might think they're having a great day with an outstanding revenue. But soon, it will return on them when payments are not fulfilled and goods are *returned to sender* as addresses are unknown.

As soon as the application and test codes reside in separate apps, the **multi-root workspace feature** of VS Code is of great help with any next coding needed to be done on them. This equips you to work in the same VS Code application window in parallel, on both apps, or maybe even a collection of multiple application apps and their test apps. As an example, you will find in the GitHub repo a folder called *LookupValue Extension* containing an *App* and a *Test* folder, that hold, respectively, the final application extension code and the final test extension code. It also holds a VS Code multi-root workspace file called `LookupValue.code-workspace`.

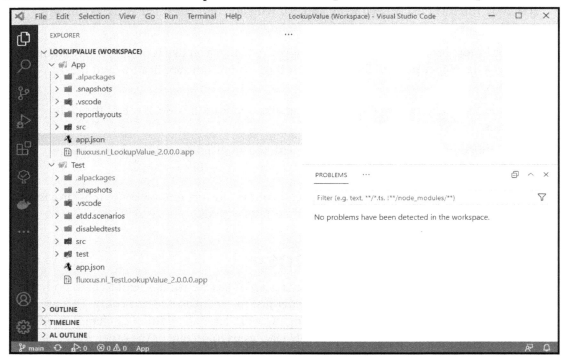

Figure 9.12 – Multi-root workspace containing both App and Test folders

> **Note**
>
> Read more about the VS Code multi-root workspace here: `https://code.visualstudio.com/docs/editor/multi-root-workspaces`.

Integrating with the daily build system

Software development is about building applications that link and automate business processes. Modern software development should do the same with respect to its own processes as follows:

- Sharing code in source code repositories that can be accessed and managed through APIs from anywhere

- Building your software from scratch at any time automatically

- Running an automated test run to show the validity of the rebuild software, triggered by a completed build, or being part of that build

- Deploying a build approved by the automated tests automatically on demand, on a scheduled time, or on every new occurrence

- Collecting all the results and statuses of the foregoing processes on a dashboard to inform the stakeholders about the health of the software

Contemporary development tools, such as Microsoft Azure DevOps, enable you to achieve this, to integrate your test automation effort with a build process. *Figure 9.13* shows a simple Azure DevOps dashboard that displays the preceding bullet points in a graphical representation:

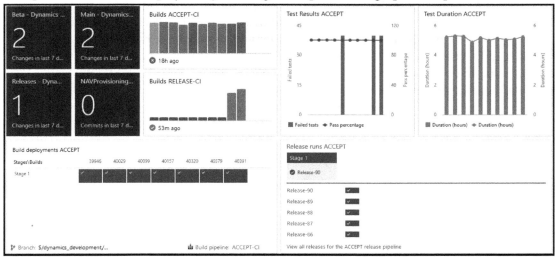

Figure 9.13 – A simple Azure DevOps dashboard

This is the world we are in with Dynamics 365 Business Central, induced by the arguments discussed in *Chapter 1, Introduction to Automated Testing*. If you're still not part of this world, make sure to get yourself moving. Not only because of the requirements Microsoft is setting for us, but – even more – because of the ecosystem our customers are living in today, where they hear about scheduled builds, automated test runs, and shorter release cycles more than ever before allowing you to go faster in adding value to their business.

Using pipelines

Continuous Integration (**CI**) and **Continuous Delivery** (**CD**) did seem for long something far out for Business Central development. With the move to AL and extension development and the use of tools like Microsoft Azure DevOps, CI/CD has become reality. Automated tests are making up an essential part of this.

Where in the last decade only a small number of Business Central development partners were putting effort in automating their development processes, more and more are picking up right now. Not in the least as the Microsoft Dynamics 365 Business Central development team started to openly advocate this too. Read, for example, the blog post series by Microsoft's Technical Evangelist Freddy Kristiansen on CI/CD: `https://freddysblog.com/category/ci-cd/`:

Figure 9.14 – A basic flowchart of a CI pipeline

A basic flow of a CI pipeline is shown in *Figure 9.14*. Given a repo containing the `.al` source code, both app, and test, the pipeline will download these, build a Docker container and build the app from scratch, publish the app to the container and run the tests, publish both app and test results, and finally, remove the Docker container. It goes without saying that you probably want your build to be reported as a failure when your tests do fail. With such a CI pipeline, aka **build pipeline**, you have a super powerful tool to inspect the health of your code, both technically and functionally, by respectively building your app(s) from scratch and running the tests that go with it.

> **Note**
>
> Might you want to read more about CI/CD and test automation from other angles, follow guys like Eric Wauters, aka waldo (`https://www.waldo.be`), James Pearson (`https://jpearson.blog/`), Michael Glue (`https://navbitsbytes.com/`), and Michael Megel (`https://never-stop-learning.de/`).

Hello World example

As part of his efforts, Freddy Kristiansen created a very nice `hello world` example of how to set up pipelines on Azure DevOps (ADO), with the use of PowerShell and Docker, to build a solution, run tests for them, and to also deploy it.

These are the direct links to the blog post and the ADO project:

- `https://freddysblog.com/2020/06/28/the-hello-world-ci-cd-sample/`

- `https://dev.azure.com/businesscentralapps/HelloWorld`

LookupValue example

For our *LookupValue* extension, I also implemented CI pipelines on ADO. For this, I did not follow Freddy's `hello world` example but instead made use of ALOps, an extension for Microsoft Azure DevOps, that facilitates you to easily set up build and release pipelines for Business Central extensions. The reason that I am using ALOps for these pipelines is twofold:

- Show you the existence of this ADO extension. You can build the pipelines entirely in ADO, but that requires you to write a lot of PowerShell yourself. ALOps provides some features that make it a lot easier to implement CI/CD specifically for BC development projects.

- Doing it the `hello world` example way does give you a lot of flexibility, but also a lot more responsibility as you need to maintain more sources, being the PowerShell scrips. Using ALOps components saves me time, as I do not need to maintain the source *behind*. Instead, I can keep my focus on building Business Central features.

> **Notes**
>
> Read more on ALOps here: `https://marketplace.visualstudio.com/items?itemName=Hodor.hodor-alops`. With it also goes a GitHub repo where very useful pipelines templates can be found: `https://github.com/HodorNV/ALOps`.
>
> Note that ALOps is not free for non-public projects.

The basis of an ADO build pipeline is a so-called **YAML** file. The pipeline example I did build for this chapter can be found here: `https://dev.azure.com/PacktPublishing/Automated-Testing-in-Microsoft-Dynamics-365-Business-Central-2Ed/_build?definitionId=5`.

This pipeline is based on this YAML file: `https://github.com/PacktPublishing/Automated-Testing-in-Microsoft-Dynamics-365-Business-Central-Second-Edition/blob/main/Chapter 09 (LookupValue Extension)/azure-pipelines.yml`

Figure 9.15 shows an overview of a run of this build pipeline:

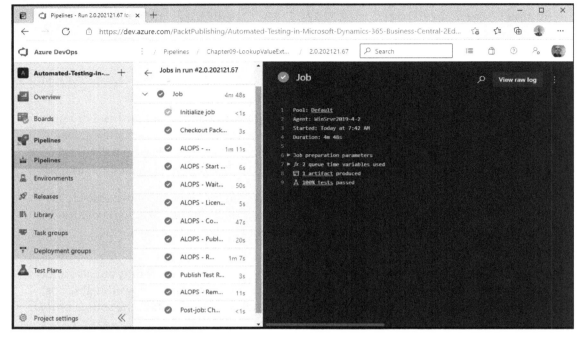

Figure 9.15 – ADO CI pipeline run for the LookupValue extension

Figure 9.16 shows the test results for the same run:

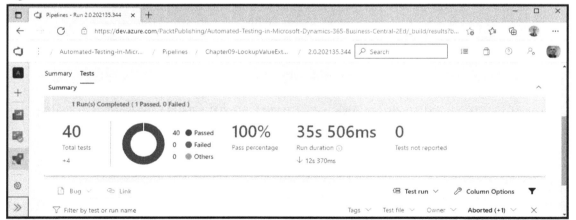

Figure 9.16 – ADO CI pipeline test results for the LookupValue extension

Similar to *Figure 9.13*, I created a simple dashboard for the *LookupValue* extension; see *Figure 9.17*. Note that I did not only build a pipeline for the code that goes with this chapter, but also for the final extension setup as discussed in the *Organizing your extensions* section. For both pipelines, you will find the `.yml` file in the respective folders on GitHub.

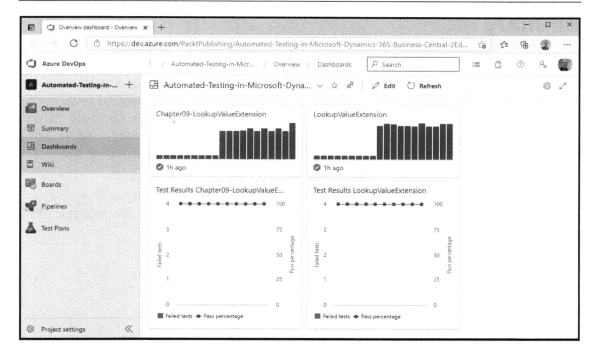

Figure 9.17 – ADO dashboard for the LookupValue extension

Notes

When you access the ADO dashboard for the LookupValue extension, you will notice that the test result widgets will not render content. This is unfortunately due to the fact that anonymous users are not granted permissions for test-related features on ADO.

Or not using pipelines

Be that as it may, you do not need to wait for fully operational automated CI/CD pipelines to get the most out of your automated tests. With a simple PowerShell script triggered by a good old Windows task, you have your tests running on your application at any scheduled time. Before the teams I worked in started implementing CI/CD pipelines on Azure DevOps, this was exactly what we had been doing.

This allowed us to execute thousands of automated tests every night with a resulting test report emailed to the team, and later in Microsoft Teams, the next morning, informing us of the health of our code. A drop in the success rate of the test run, which happened occasionally, notified us that some unintended breaking changes had been added to our application and appropriate actions were needed.

Summary

In this chapter, we paid attention to a number of best practices on how to embed test automation in your daily development practice. Get your functional peers writing ATDD scenarios to make use of the Excel sheet discussed. Do not overload yourself and your team and take small steps. Use the test tool next to your development tools and keep the tests running. Be aware that test code is also code, so maintain it as you maintain your application code. Organize your final application app, and its accompanying test app, in such a way that tests cannot be run in a production environment accidentally. And last but not least, automate parts of your development processes with automated builds that include running your tests.

In the next chapter, *Chapter 10, Getting Business Central Standard Tests Working on Your Code*, we will have a closer look at the tests provided by Microsoft and how we could also integrate this humongous collateral of standard tests.

10
Getting Business Central Standard Tests Working on Your Code

By now, you know how to write your automated tests and have integrated them in your daily development practice. How can you profit from this humongous collateral of tests Microsoft has provided? This chapter discusses why you might want to add them to your own collateral, what hurdles you must take, and how. So, in this chapter we'll discuss:

- Why use the standard tests?
- Executing standard tests against your code
- Fixing failing standard tests
- Disabling failing tests
- It's all about data
- Is it all really about data?

Technical requirements

Like in the previous chapter, we will refer to the **LookupValue** extension. Its code is to be found on GitHub: `https://github.com/PacktPublishing/Automated-Testing-in-Microsoft-Dynamics-365-Business-Central-Second-Edition`.

In this chapter, we will continue on the code from *Chapter 9, How to Integrate Test Automation in Daily Development Practice*: `https://github.com/PacktPublishing/Automated-Testing-in-Microsoft-Dynamics-365-Business-Central-Second-Edition/tree/main/Chapter 09 (LookupValue Extension)`. In the current chapter – *Chapter 10* – we will only focus on the Business Central standard tests. We can disregard our own test codeunits, and therefore we can throw them out. We will, however, retain our test libraries.

The final code of this chapter can be found in `https://github.com/PacktPublishing/Automated-Testing-in-Microsoft-Dynamics-365-Business-Central-Second-Edition/tree/main/Chapter 10 (LookupValue Extension)`.

Details on how to use this repository and how to set up VS Code are discussed in *Appendix, Getting Up and Running with Business Central, VS Code, and the GitHub Project*.

Why use the standard tests?

Ever since they introduced the testability framework in 2009, Microsoft has been building on its application test collateral. As already pointed out in *Chapter 4, The Test Tool, Standard Tests, and Standard Test Libraries*, it contains an immense number of tests. These tests cover the whole standard application, from financial management, sales, and purchase, through warehouse and manufacturing, to service management. With every major release or cumulative update, new tests have been, and will continue to be, added to cover new features and recent bug fixes. Years of work we all can profit from. If your code extends the standard application, what will the impact be on it?

You could go about writing your own tests. You could also choose to run the standard tests and see the results. And, of course, in the end, you could do both, as your extension most probably will not only change standard behavior but also add new functionality not covered by any test in the Microsoft collateral.

We could discuss a lot and bang each other's heads together on the validity of running the Microsoft tests. Instead, we could just as well run them and see if our code makes the standard tests fail. This is exactly what I did years ago as my first step in structurally picking up test automation for Business Central.

Those that have been following my blog might recall that some years ago, I dedicated a post on this called *How-to: Run Standard Tests against Your Code*: `https://www.fluxxus.nl/index.php/BC/how-to-run-standard-tests-against-your-code`. As the title unveils, it discusses the steps to take to get the standard tests run on your solution. Within 30 minutes, you can have the tests running, and within a couple of hours, you will have the results. The chart in *Figure 10.1* displays the result of a full standard test collateral run on a solution I was working on at my former employers, back in 2016, on NAV 2016:

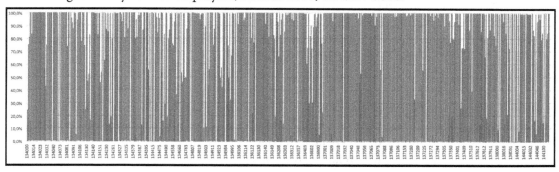

Figure 10.1 – Graphical representation of the first test run of standard tests on our solution (blue, darker spikes for success percentage and orange, lighter spikes for failure percentage)

The *blue*, darker spikes show the success percentage of each test codeunit, while the *orange*, lighter spikes account for the failure percentage for each test codeunit. Overall, only approximately 23% of the standard tests were successful. When I first digested this result, it felt a bit disappointing, but actually, it was a very good result as this meant that approximately 77% of the standard tests did in one way or another hit – some of – our code, which made these tests fail. This test, run on NAV 2016, showed us that out of a total of 16,124 standard tests, almost 12,450 were touching our code. If we could get these to succeed, we would have started with a substantial collateral of automated tests. In only 5 days we lifted the success rate from approximately 23% to 72%. Another work week raised it by almost 80%. We stopped at 90% another 4 weeks later. Six weeks of work gave us the availability over approximately 10,800 tests that triggered – some of – our code.

Imagine having to create that many tests yourself. A quick calculation gives us an impression: let's assume optimistically that you would be able to write one automated test per 10 minutes. This comes down to a total of 1,800 hours, which is in the order of one year of work for one person. Mind you, these 6 weeks we invested weren't that bad. And you know the 23% successful tests also come in handy, as by running these we safeguard that we do not compromise the behavior of the code they test.

Executing standard tests against your code

The proof of the pudding is in the eating, so you can go and verify it by following these steps:

1. Deploy the solution we built and tested in *Section 3, Designing and Building Automated Tests for Microsoft Dynamics 365 Business Central*, to Business Central.

2. Deploy all standard test extensions and test libraries – see *Getting standard test and test library apps deployed* in *Appendix, Getting Up and Running with Business Central, VS Code, and the GitHub Project*.

3. Set up a test suite in the test tool with all present test codeunits, as discussed in *Chapter 4, The Test Tools, Standard Tests, and Standard Test Libraries*.

4. Run all the tests.

This worked pretty fine on C/SIDE-based versions, but unfortunately, moving forward to newer versions that are AL-based with more and more tests added, the standard test collateral showed to be unstable to some extent. While a standard test run on a NAV 2018 CRONUS, demo company would yield only a few dozen failing tests on a total of over 20,000 tests, a standard test run on BC17.4 produces about 3,500 erroneous tests on a total of over 35,000 tests.

In early 2021, I did a project for a Business Central implementation partner to get the standard tests running on their solution. Then I encountered different aspects of the instability of these tests, such as the following two:

*   Only having the Tests-ERM test extension installed – not run – next to all other standard test apps made up for over 1,100 failing tests out of the total of 3,500 failing tests. The consequence was that you could not execute all standard tests in one run. Uninstalling the Tests-ERM test extension would let those 1,100 succeed.

*   Running the tests of a standard test extension for the first time would show that a certain number did not succeed. After disabling these tests, rerunning them would often show other tests failed. The Tests-SCM test extension proved to be a difficult-to-tame beast as it took 22 runs to filter out all the failing tests. The challenge was not only the number of runs but also the length of a run being over 4 hours.

> **Note**
> The fact that we find certain tests not passing, a second and another run reveals that some tests appear to have, often unintended, dependencies. Data created in one test is expected in the next test. If the first test is disabled, its data is not created, and the second test will fail due to missing data. It's another kind of **test smell**.

Many Business Central peers have validly been complaining about this instability and have asked Microsoft to solve it, such as Huub van Hout did with his idea on *Dynamics 365 Application Ideas*: `https://experience.dynamics.com/ideas/idea/?ideaid=074896c9-e211-ea11-b265-0003ff68f60b`.

Executing Tests-VAT

Due to the instability mentioned in the previous section, you cannot run all standard tests in one go but, instead, install and run one test extension after the other. For this, we now pick one representative test extension: `Tests-VAT`. As the name suggests, this extension contains tests that check the VAT features of Business Central, altogether over 900 tests. Your execution plan becomes:

1. Deploy the solution you built to Business Central.

2. Deploy the selected standard test extension, in this case, `Tests-VAT` and test libraries.

3. Set up a test suite in the test tool with all present test codeunits.

4. Run all the tests.

Depending on the standard test extension you deploy, and the resources of the machine you run the tests on, the test run might be a matter of minutes up to an hour, or even a few hours. In the case of the `Tests-VAT` extension, the execution time is just over 10 minutes with a result of 760 successful and 175 failing tests.

> **Note**
>
> For the project with the aforementioned Business Central implementation partner, I ran all the tests pipelines on an Azure VM with specs *Standard E4as_v4 (4 vcpus, 32 GiB memory)*. Based on the test run duration, I grouped the tests apps as follows:
>
> a) *Up to 5 minutes*: Tests-Cash Flow / Tests-Cost Accounting / Tests-Data Exchange / Tests-Fixed Asset / Tests-Local / Tests-Monitor Sensitive Fields / Tests-Permissions / Tests-Physical Inventory / Tests-Resource / Tests-Reverse / Tests-SMTP / Tests-Upgrade.
>
> b) *Up to 10 minutes*: Tests-Bank / Tests-General Journal / Tests-Graph / Tests-Integration / Tests-Invoicing / Tests-Job / Tests-Marketing / Tests-Prepayment / Tests-Rapid Start / Tests-User.
>
> c) *Up to 20 minutes*: Tests-CRM integration / Tests-Dimension / Tests-Report / Tests-VAT.
>
> d) *Up to 1 hour*: Tests-Misc / Tests-SMB / Tests-Workflow.
>
> e) *Beyond 1 hour*: Tests-ERM / Tests-SCM.
>
> Of course, having a different machine setup will possibly speed up your test run.

What does this tell us?

First of all, 175 from a total of 935 tests is quite a substantial number of failures. Secondly, from experience, not even looking at the error text, I can tell this is mainly related to our extension for the following reason: running `Tests-VAT` tests on standard Business Central throws about 9 errors. Now, doing a simple calculation, this means that at least approximately 166 of the `Tests-VAT` tests do hit our code.

Only selecting tests from Tests-VAT

If you want to run only the tests from the `Tests-VAT` test extension you have the following options:

- If you want to run the tests in the test tool, make sure you only have `Tests-VAT` installed – and the standard test libraries – and no other test extensions.

  Use **Get Tests Codeunits by Range** – see *Chapter 4, The Test Tools, Standard Tests, and Standard Test Libraries* – and set the filter to **only select tests codeunits** above **134011** (`Sys. Warmup Scenarios`). This will get all test codeunits from the `Tests-VAT` test extension.

- Using `Run-TestsInBcContainer` from the **BcContainerHelper** PowerShell module with the parameter `extensionId` set to the ID of the `Tests-VAT` test extension. You can either use this in a script you trigger manually or in a pipeline. ALOps provides something similar.

  In this blog post, you can read more about using `Run-TestsInBcContainer`: `https://freddysblog.com/2019/10/22/running-tests-in-containers-2/`.

This applies, of course, to any other test extension. Later in this chapter, we employ it for `Tests-Fixed Asset`.

> **Note**
> Unfortunately, there is no explicit way of knowing what codeunits belong to what test extension. To get explicit feedback on the test results of one test extension, you have to install and run the test for one test extension at a time. Depending on your setup, this might mean that for each run you recreate the environment – when using, for example, a Docker container – or uninstall the previous test extension first.

Running tests using pipelines

Two pipelines have been added to the same Azure DevOps project that holds the pipeline discussed in *Chapter 9*, *How to Integrate Test Automation in Daily Development Practice*:

- **Chapter10-BC18-Tests-VAT** to run the `Test-VAT` tests against the Business Central base application – aka BaseApp

- **Chapter10-LookupValueExtension-BC18-Tests-VAT** to run the `Test-VAT` against our *LookupValue* extension – deployed on the BaseApp

Figure 10.2 shows the dashboard that goes with the two pipelines displaying from left to right the build results, the test run results, and the test run durations:

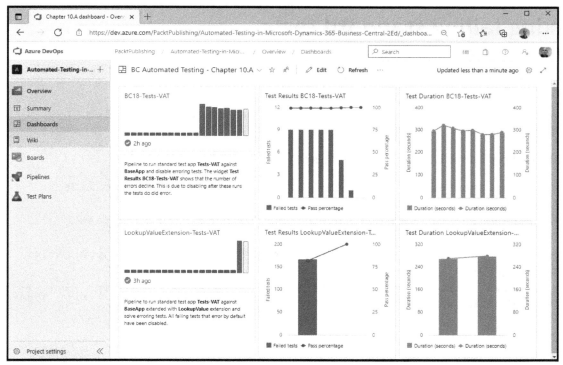

Figure 10.2 – ADO dashboard for pipelines to run Tests-VAT

Follow this link to have a look for yourself: `https://dev.azure.com/ PacktPublishing/Automated-Testing-in-Microsoft-Dynamics-365- Business-Central-2Ed/_dashboards/dashboard/654b1f68-fa79-4cc2-a279- 9e3d78e63ff5.`

> **Notes**
>
> (1) As mentioned in *Chapter 9, How to Integrate Test Automation in Daily Development Practice*, accessing the ADO dashboard anonymous users, the test result widgets will unfortunately not render content.
>
> (2) The YAML file defining the pipeline discussed earlier can be found in the GitHub folder for *Chapter 10* and is called `azure-pipelines-Tests-VAT.yml`.
>
> (3) When writing the book, the pipelines were only run against Business Central 2021 wave 1, aka B18, as the wave 2 version, BC19, was still in development. Once this version is released, I will make sure the ADO project also contains pipelines that run against it.

Fixing failing standard tests

Now let's have a look at the errors and see which of them relate to the extension. They might already unlock some of their secrets. All errors that relate, about 157 out of the 176, are similar to the following:

```
Error Message: Lookup Value Code must have a value in Sales
Header: Document Type=Order, No.=1003. It cannot be zero or empty.
```

With also an **Error Call Stack** that resembles this one:

```
SalesPostEvents(CodeUnit 50004).OnBeforePostSalesDocEvent line 3 -
LookupValue by fluxxus.nl
"Sales-Post"(CodeUnit 80).OnBeforePostSalesDoc line 2 - Base
Application by Microsoft
"Sales-Post"(CodeUnit 80).OnRun(Trigger) line 26 - Base Application
by Microsoft
"Library - Sales"(CodeUnit 130509).DoPostSalesDocument line 53 -
Tests-TestLibraries by Microsoft
"Library - Sales"(CodeUnit 130509).PostSalesDocument line 2 -
Tests-TestLibraries by Microsoft
"ERM VAT Tool - Sales Doc"(CodeUnit 134051).
VATToolSalesLineWithZeroOutstandingQty line 14 - Tests-VAT by
Microsoft
"Test Runner - Mgt"(CodeUnit 130454).RunTests - Test Runner by
Microsoft
"Test Runner - Isol. Codeunit"(CodeUnit 130450).OnRun - Test
Runner by Microsoft
```

```
"Test Suite Mgt."(CodeUnit 130456).RunTests - Test Runner by
Microsoft
"Test Suite Mgt."(CodeUnit 130456).RunTestSuiteSelection - Test
Runner by Microsoft
"AL Test Tool"(Page 130451)."RunTests - OnAction" - Test Runner by
Microsoft
```

For those of you with some years of experience as a developer in the Business Central world, the format of the error message may already hint at its cause: an error being thrown at the record method `TestField`.

Now let's handle the errors. This is what I call my *attack protocol*, which you perform after you have executed the relevant standard tests:

1. From VS Code, start up a debug session.
2. Open the Test Tool and select one of the failing tests.
3. Run the individual test using **Run Selected Tests**.
4. Let the debugger break on the error.
5. See where the error occurred.
6. Use the **call stack** to step back in the code and see if you can distinguish the cause, or if you want to get some more details.
7. Place a breakpoint somewhat earlier in the code.
8. Finish the code execution and rerun the test.
9. Debug by stepping through the code.
10. Implement the fix and restart from *step 1*.

Ready? Follow me in attacking the error!

> **Notes**
>
> (1) The numbers I am using in this chapter might be somewhat different when you are reproducing my scenarios using a different build or minor/major version of Business Central.
>
> (2) To get the different numbers, I used the **Open in Excel** action on the Test Tool and filtered the rows in the resulting Excel sheet.

Attacking the error

Continuing with the code from *Chapter 9, How to Integrate Test Automation in Daily Development Practice*, let's follow the steps of my *attack protocol*:

1. From VS Code, start up a debug session.

2. Open the test tool and select the failing test.

3. In our case, pick any test that fails with the `Lookup Value Code must have a value...` error:

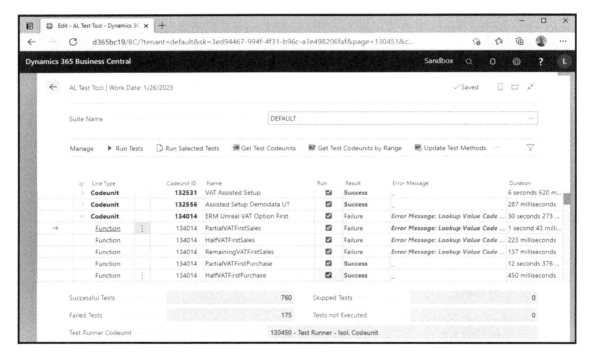

Figure 10.3 – Open the test tool and select the failing test

4. Run the individual test using **Run Selected Tests**:

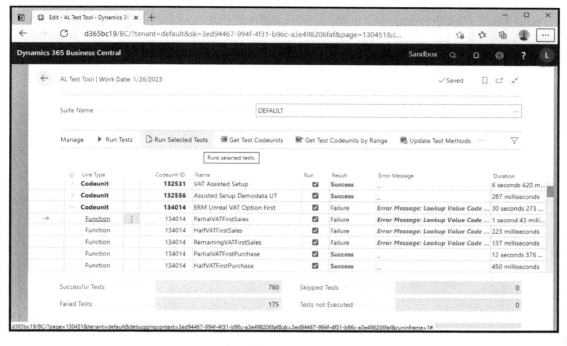

Figure 10.4 – Run the individual test using Run Selected Tests

5. Let the debugger break on the error:

Figure 10.5 – Let the debugger break on the error

6. See where the error occurred. As can be seen in *Figure 10.5*, it happened in an object in our extension. It might, however, not be familiar to you as we haven't had a look at this code yet. It's the implementation of the business rule that says that when posting a sales document, it is mandatory that the `Lookup Value Code` field is populated. See the call to the `TestField` method.

7. Clearly, somewhere back in the execution path, the `Lookup Value Code` field did not get a value assigned.

8. Use the call stack to step back in the code. Notice that I selected the line on the call stack just above the line with `RunTest`. `RunTest` is the main function in the standard test runner codeunit (130450) that calls each test codeunit. The line on top of that is always the call to the current test function. In this specific case, `PartialVATFirstSales` as shown in *Figure 10.6*:

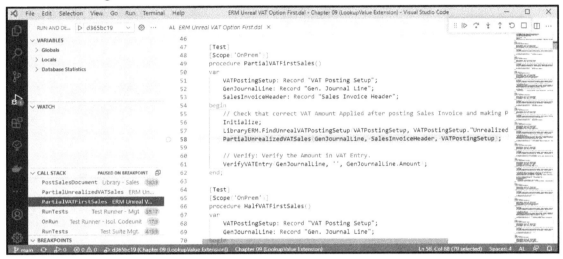

Figure 10.6 – Use the call stack to step back in the code

`PartialVATFirstSales` is our starting point to find out where we can hook in to get the error fixed, which ensures that the `Lookup Value Code` field is populated. My first impression tells me to take a closer look at `Initialize` and `PartialUnrealizedVATSales`, and see if it creates and uses a customer on the sales header, or that we need to populate the `Lookup Value Code` field on the sales document itself:

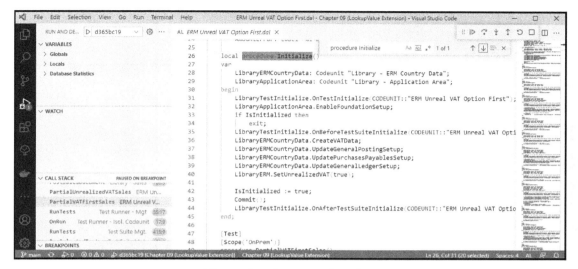

Figure 10.7 – Inspecting Initialize

As also shown in *Figure 10.7*, `Initialize` doesn't really seem to do anything with customer or sales document data. So, how about `PartialUnrealizedVATSales`?

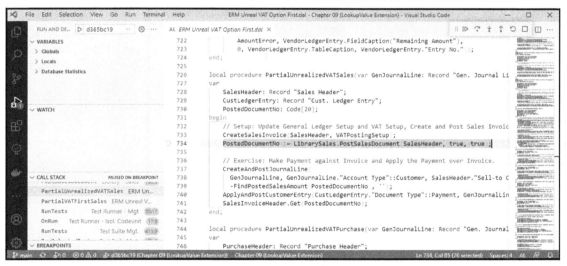

Figure 10.8 – Inspecting PartialUnrealizedVATSales

As the selected statement in *Figure 10.8*, part of our execution path, will post a sales document it most probably is the previous statement, which calls `CreateSalesInvoice`, we need to step in to investigate more:

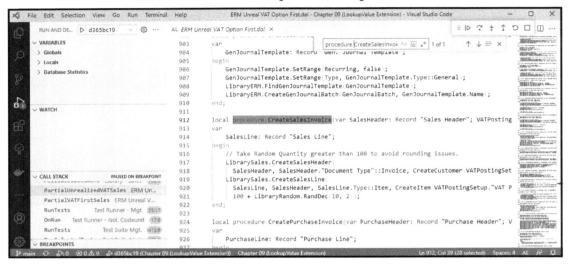

Figure 10.9 – Inspecting CreateSalesInvoice

Do you see what I see in *Figure 10.9*? Line 918! `CreateCustomer`! Apparently, a local method exists that creates a customer as part of the *fresh fixture* (*Figure 10.10*). I would bet my money – all my money – that this customer did not get a `Lookup Value` assigned. To find out, let's execute the next step of our *attack protocol*:

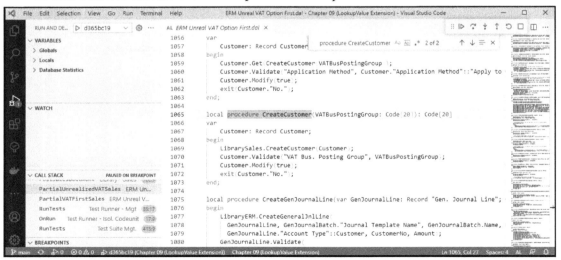

Figure 10.10 – Inspecting CreateCustomer

9. Place a breakpoint somewhere earlier in the code. You might have guessed where already, at the line calling `CreateSalesInvoice`:

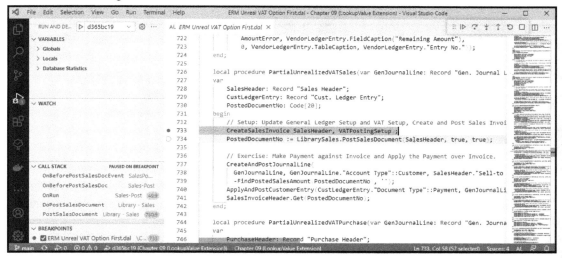

Figure 10.11 – Place a breakpoint somewhere earlier in the code

10. Finish the code execution and rerun the test. As expected and shown in *Figure 10.12*, it stops at the `CreateSalesInvoice` call:

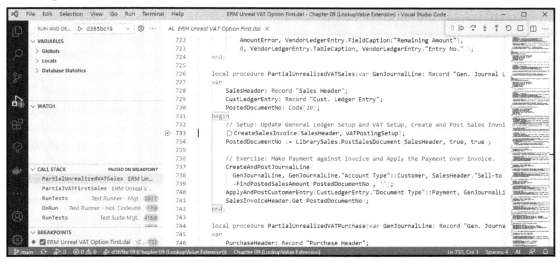

Figure 10.12 – Finish the code execution and rerun the test

11. Debug by stepping through the code. Now, one after the other, we first step into `CreateSalesInvoice`, then into the local `CreateCustomer` method that is called by `LibrarySales.CreateSalesHeader`:

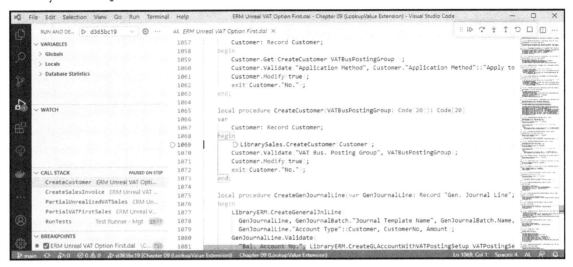

Figure 10.13 – Stepped into CreateCustomer

As can be seen in *Figure 10.13*, the local helper function `CreateCustomer` calls the standard helper method `CreateCustomer` in the `Library - Sales` codeunit:

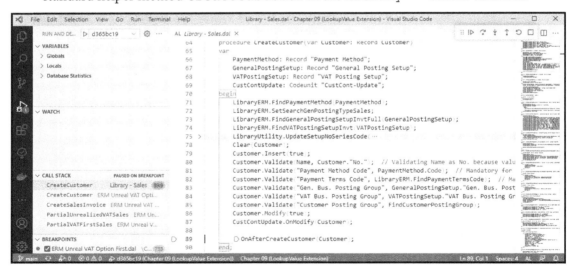

Figure 10.14 – Stop at OnAfterCreateCustomer

Right here is where we would like to fix the issue, not least because this helper function typically will be called from almost any test that is needing a newly created customer. Have you noticed the OnAfterCreateCustomer publisher? Our fix is going to subscribe to it.

12. Implement the fix and restart from *step 1*.

We will elaborate on the last step in the next section.

> **Note**
>
> Microsoft has started to add publishers such as OnAfterCreateCustomer to helper functions in their libraries to let us extend them. You might, however, still run into quite a few helpers that have not yet been blessed with a publisher. It is possible to request for them, like I did with the ones I needed. Unfortunately, there is no official route for this apart from the *Test Tool* group on the *Dynamics 365 Business Central Development* Yammer group: https://www.yammer.com/dynamicsnavdev/#/threads/inGroup?type=in_group&feedId=13879726.

Fixing the error

To fix the error, the trick is to add a codeunit to our extension with a subscriber to the OnAfterCreateCustomer publisher that sets a lookup value on the customer:

```
codeunit 80050 "Library - Sales Events"
{
    [EventSubscriber(ObjectType::Codeunit,
        Codeunit::"Library - Sales", 'OnAfterCreateCustomer',

        '',false, false)]
    local procedure OnAfterCreateCustomerEvent(
                var Customer: Record Customer)
    begin
        SetLookupValueOnCustomer(Customer);
    end;

    local procedure SetLookupValueOnCustomer(
                var Customer: record Customer)
```

```
var
  LibraryLookupValue: Codeunit "Library - Lookup Value";
begin
  Customer.Validate("Lookup Value Code",
    LibraryLookupValue.CreateLookupValueCode());
  Customer.Modify();
end;
}
```

Note the reference to the codeunit `Library - Lookup Value`. You might recall this is the result of refactoring in *Chapter 8, From Customer Wish to Test Automation – the TDD way*. `Library - Lookup Value` contains a reusable function called `CreateLookupValueCode`.

Running the failing tests again

Deploy everything again and rerun all the failing tests. Use the **Select on Failures** feature from our Test Tool extension to select only the failing tests, and then run all, being only the previously failing tests. The result will be that the number of failing tests has declined to only about two handfuls. Clearly, this fix was a good investment.

> **Note**
>
> Typically, when getting failing standard tests fixed that were run against your code, the debugger is an indispensable tool. If you want to see me use the debugger, and some more, you might want to watch the Areopa webinar *How about verifying and debugging your tests* on YouTube: `https://www.youtube.com/watch?v=zna9iKllz-o`.
>
> In *Chapter 11, How to Construct Complex Scenarios*, we'll do some more debugging.

Disabling failing tests

How about the almost two handfuls of remaining failing tests? These are standard tests that at this moment do not succeed in your environment. Whether it is because they are failing anyway or because they need some specific setup, it's not clear to me. In general, I will just disable them on the assumption that it's a relatively and absolutely small number of tests, and that the remaining successful ones give me a good enough health check of the code. Of course, if time allows, you might want to study them closer, but if the numbers are so low, please ask yourself if spending more time is worthwhile.

Disabling failing tests could be done manually using, for example, the test tool extension. A more repeatable solution, however, would be to use the **DisabledTests.json** feature provided by the Run-TestsInBcContainer cmdlet of the BcContainerHelper module, or one of the ALOps pipeline steps. To be able to create a valid DisabledTests.json file, you should run standard tests on CRONUS to determine what tests do fail. Based on this outcome you then create the .json file. For the Tests-VAT test app this leads to the following content, showing only the first and last entry:

```
[
  {
    "codeunitId": 134025,
    "codeunitName": "ERM Unrealized VAT Customer",
    "method": "ApplyPaymentToInvoice"
  },

  ...

  {
    "codeunitId": 135401,
    "codeunitName": "Fixed Assets Plan-based E2E",
    "method": "FixedAssetJourneyAsExternalAccountant"
  }
]
```

Go to GitHub to set eyes on the complete set and notice that this file contains more than the two handfuls of remaining failing tests. As already discussed with respect to the Tests-SCM test app, this is due to the fact that when you disable these and rerun all Tests-VAT tests, a number of other tests fail. Three test runs were needed to find all the failing tests listed earlier.

Notes

(1) You can find the DisabledTests.json file that I created in the GitHub repository.

(2) Given the test result in a so-called TestResults.xml, an output of Run-TestsInBcContainer, or one of the ALOps pipeline steps, you can use a script from Freddy Kristiansen to generate a DisabledTests.json. Find the script here: https://github.com/microsoft/navcontainerhelper/issues/748#issuecomment-564728120.

(3) Also, have a look at the pipelines that go with this chapter and include a step that generates DisabledTests.json and attaches it as an artifact to the pipeline run.

Syntax of DisabledTests.json

In `DisabledTests.json`, you specify either a single test function or a whole test codeunit that you want to be disabled and the minimum info needed is:

- For a single test function:

  ```
  { "codeunitName": "<codeunit name>", "method": "<method name>"
  }
  ```

- For a whole test codeunit:

  ```
  { "codeunitName": "<codeunit name>", "method": "*" }
  ```

The `codeunitId` keyword is optional, but I always try to add it as it is often a simpler reference for developers.

It's all about data

In my experience, getting the standard tests working on your code is mainly about getting the test fixture right, as in the previous exercise. Fixing the error by supplementing the fixture isn't a coincidence. It's the most probable thing that you will be doing in getting the standard tests run on your code: bring the test fixture in the right state.

In this specific case, we fixed the *fresh fixture*. Running more standard tests will show other kinds of test errors. Solving these errors, however, is again mostly about updating the test fixture, *fresh* and, in some cases, *shared fixture*.

Executing and fixing Tests-Fixed Asset

Having run only one standard test app – `Tests-VAT` – let's see the results of running another one against our code: `Tests-Fixed Asset`. Some of those failures relate to the *LookupValue* extension. Apparently, the fix we implemented to tackle the errors thrown on the `Tests-VAT` test run did not prevent the current errors from happening in `Test-Fixed Asset`.

Let's focus on one example and discuss the fix for it. *Figure 10.15* shows the error that occurs for test function `PartialDisposalOfFA` in test codeunit 134450 (ERM Fixed Assets Journal):

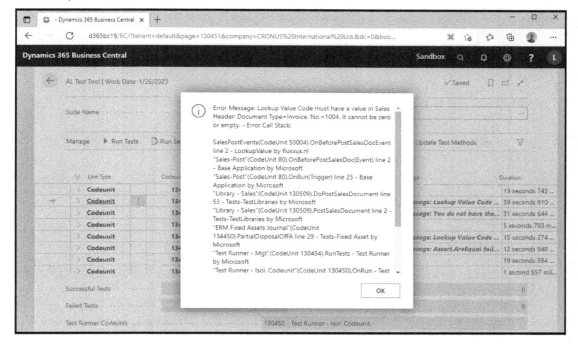

Figure 10.15 – Error of failing test PartialDisposalOfFA

Even though the error is identical to the ones we did fix, clearly our fix for the standard `CreateCustomer` helper method did not have any effect on this.

Debugging `PartialDisposalOfFA`, following my *attack* protocol, uncovers that `Test-Fixed Asset` does not use the `CreateCustomer` library function. Instead, prebuilt fixture, that is, data from the CRONUS demo company, is being used, as *Figure 10.16* shows us:

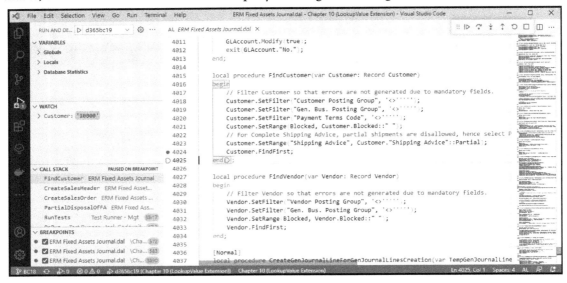

Figure 10.16 – Local helper function FindCustomer retrieving an existing customer

Examining the code of the `FindCustomer` local helper function, we learn that within a set of filters, the first customer record is retrieved from the database. Reading from the **Watch** pane, it turns out to be a well-known customer in CRONUS, customer `10000`. As we haven't touched the prebuilt fixture, indeed this record, like any customer record in CRONUS, must do without a lookup value.

To fix this error we have to update the prebuilt fixture, that is, create a *shared fixture*, making sure that customers already residing in CRONUS get their `Lookup Value Code` field populated. To achieve this, we can make use of one of the publishers in the shared fixture method, `Initialize`, implemented in most of the Microsoft test functions. You might recall this structure from *Chapter 5*, *Test Plan and Test Design*:

```
local Initialize()
// Generic Fresh Setup
LibraryTestInitialize.OnTestInitialize(<codeunit id>);
<generic fresh data initialization>

// Lazy Setup
if isInitialized then
  exit();
```

```
LibraryTestInitialize.OnBeforeTestSuiteInitialize(<codeunit id>);

<shared data initialization>

Initialized := true;
Commit();

LibraryTestInitialize.OnAfterTestSuiteInitialize(<codeunit id>);
```

We subscribe to either the OnBeforeTestSuiteInitialize or
OnAfterTestSuiteInitialize publisher. In general, I chose to subscribe to the first one
to make sure this is done before any standard update to the fixture is performed and to make use
of the already present call on Commit.

This is how our fix implementation looks:

```
codeunit 80051 "Library - Initialize"
{
  [EventSubscriber(ObjectType::Codeunit,
    Codeunit::"Library - Test Initialize",
    'OnBeforeTestSuiteInitialize','', false, false)]
  local procedure OnBeforeTestSuiteInitializeEvent(
      CallerCodeunitID: Integer)
  begin
    Initialize(CallerCodeunitID);
  end;

  local procedure Initialize(CallerCodeunitID: Integer)
  var
    LibraryLookupValue: Codeunit "Library - Lookup Value";
    LibrarySetup: Codeunit "Library - Setup";
  begin
    case CallerCodeunitID of
      Codeunit::"ERM Fixed Assets Journal",
      Codeunit::"ERM Fixed Assets GL Journal":
        LibrarySetup.UpdateCustomers(
          LibraryLookupValue.CreateLookupValueCode());
    end;
  end;
}
```

Notice this code also fixes the same error thrown by tests in codeunit `134453` (ERM Fixed Assets GL Journal) by updating the existing records in the `Customer` table.

The newly introduced helper `UpdateCustomers` function in the new library codeunit `Library - Setup` does exactly what its name describes: it updates all customer records already existing in CRONUS so their `Lookup Value Code` field is populated.

You might argue that it would suffice and be a nice and efficient *fake it* in the spirit of TDD to only update customer `10000`. True. In the same spirit of TDD, fixing standard tests should be simple and fast. Updating all existing customer records is in line with that.

Refer to GitHub to study the details of each of these functions.

Running tests using pipelines

Like for `Tests-VAT`, I have created similar pipelines to run the tests of `Tests-Fixed Asset` and a dashboard to display the results (see *Figure 10.17*).

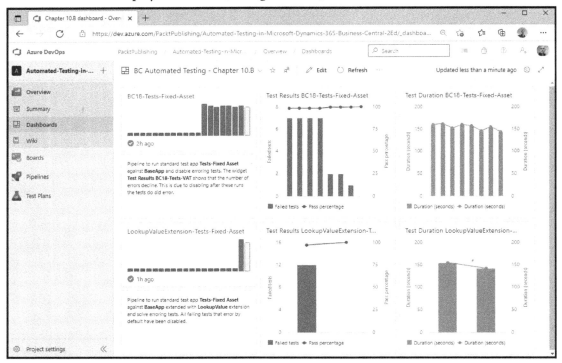

Figure 10.17 – ADO dashboard for pipelines to run Tests-Fixed Asset

Follow this link to have a look for yourself and select dashboard *BC Automated Testing – Chapter 10.B*: `https://dev.azure.com/PacktPublishing/Automated-Testing-in-Microsoft-Dynamics-365-Business-Central-2Ed/_dashboards/dashboard/4a458b07-931b-458e-b322-c62a72b4e6f7`.

Is it all really about data?

Of course not! But in the majority of the cases of failing standard tests, it worked out that way for the projects I was on. Based on this experience, my best practice in getting a failing standard test fixed is the following sequence:

1. Try fixing the error by adjusting the shared fixture, if that does not do the job adjust…

2. The fresh fixture, if not adjust…

3. The test code, often the verification part. In this case, you hit upon a test code bug. If it is still not working, adjust…

4. The application code. You found a *real* bug!

This helped me a lot in having the right focus in getting standard tests working on our solution.

Summary

In this chapter, you learned why you would want to use the Microsoft tests, how to start applying them on your Business Central code, and how to fix the most regularly occurring errors they might throw. You learned that most fixes relate to adjusting the test fixture, be it *shared* or *fresh*.

With this chapter, the first edition of the book came to an end. In this second edition, we have added a whole new section, *Section 5, Advanced Topics*, containing three new chapters. There, we will discuss how to construct complex scenarios, write testable code, and test outgoing and incoming calls and how to mock them.

Section 5: Advanced Topics

Now that you know the basis of test automation and have been given a number of tips on how to incorporate test automation into your daily practice, we elaborate in this section on a number of useful, advanced topics.

This section contains the following chapters:

- *Chapter 11, How to Construct Complex Scenarios*
- *Chapter 12, Writing Testable Code*
- *Chapter 13, Testing Incoming and Outgoing Calls*

11

How to Construct Complex Scenarios

Whereas the examples in *Section 3, Designing and Building Automated Test for Microsoft Dynamics 365 Business Central*, covered the main, one-fold principles of an automated test using the ATDD format, this chapter shows you how to go about complex scenarios.

To achieve your goals as an end-user in Dynamics 365 Business Central, being the ERP package it is, you often end up exercising a substantial chain of consecutive actions. Say you need to test applying a payment to a posted invoice. In order to run this process, a number of related things need to happen first. You have to create a customer, create an item, create and post the invoice, and create and post the payment. And eventually, verify all relevant results.

How do you go about constructing a suite of tests for that? How to create reusable parts? And how to make use of already existing helper functions in the Microsoft test libraries?

In this chapter, we will get answers to these questions when discussing the following:

- What makes a scenario complex?
- Composing complex scenarios
- Creating reusable parts
- Test example 10 – how to construct an extensive scenario

Technical requirements

Like in previous chapters, we will refer to the **LookupValue** extension. Its code is to be found on GitHub: `https://github.com/PacktPublishing/Automated-Testing-in-Microsoft-Dynamics-365-Business-Central-Second-Edition`. In this chapter, we will continue on the code from *Chapter 9*, *How to Integrate Test Automation in Daily Development Practice*, and not include our findings from *Chapter 10*, *Getting Business Central Standard Tests Working on Your Code*: `https://github.com/PacktPublishing/Automated-Testing-in-Microsoft-Dynamics-365-Business-Central-Second-Edition/tree/main/Chapter 09 (LookupValue Extension)`.

The final code of this chapter can be found in `https://github.com/PacktPublishing/Automated-Testing-in-Microsoft-Dynamics-365-Business-Central-Second-Edition/tree/main/Chapter 11 (LookupValue Extension)`.

Details on how to use this repository and how to set up VS Code are discussed in *Appendix*, *Getting Up and Running with Business Central, VS Code, and the GitHub Project*.

What makes a scenario complex?

Complexity is a relative concept. It highly correlates with the perspective and expertise of the person or team facing it. Nevertheless, there are generic, common facets that can introduce complexity. In the context of test automation, complexity can be found in the following factors:

- Data setup
- Verification
- End to end code execution path
- Dependencies on other systems

Recognizing each of these factors in general terms – and, of course, even more in specific circumstances – helps you a lot in reducing the complexity. Or should I say: *handling the complexity*? It will support you to deconstruct a complex issue into a collection of simpler things. Even though this collection may still be overwhelming as a whole, it enables you **to solve it step-by-step**, a recurring approach in this book. An approach that allows you to take small steps and gives you a great chance to design *reusable*, *readable*, and *minimalistic* code.

Let's have a closer look at each of the aforementioned listed factors and determine how complexity might come into play. After that, we will elaborate on how to compose complex scenarios in the next section.

Data setup

Getting the fixture for a test in place might be obvious and it might not take much time. But more often than not, it consumes a substantial part of your effort to get a test executed – manually – or coded. Based on my experience I tend to say that, in both cases, getting a fixture in place is about half of the effort.

The complexity of your data setup lies in the amount and/or kind of data needed to be able to exercise your test. Or, maybe more: it lies in the ignorance of the extent of the data setup needed for your test. For many of us, our testing experience is a manual one. This manual testing is mostly done on existing and regularly used databases. As such, manual testing relies – at least partly – on data already present. Creating, for example, a new master data record for your manual test seems a simple action. Not a lot of data is involved as you might need to provide some address information, payment terms, and that's it. In the background, however, creating a master data record relies, for example, on the fact that a number series has been set up, a posting setup exists, and address integration-related matters are present.

As discussed in *Chapter 5, Test Plan and Test Design*, automated tests have to be reproducible and therefore do not rely on data residing in the database. In other words, automated tests should be **data agnostic**, meaning that they should not rely on data already present, but instead create it on the fly. The consequence is that your data setup – the `GIVEN` part(s) in your **ATDD** scenario – has to take into account all the data needed for your test, either as fresh or shared fixture: see the examples in previous chapters.

Indeed, getting all the details of an extended data setup in place can be tedious. But the more you work on this, the more efficient and effective you get. On one hand, you will know how to efficiently and effectively design and on the other how to reuse data setup in the same project, but also across different projects. Or like I phrased it in *Chapter 5, Test Plan and Test Design*, as a team you will find out and learn what is enough to keep it *simple and yet thorough*.

Notes

(1) My rough relative time estimation for getting the test data setup is 50% to 60% of the overall work on tests. The more test libraries you have built with helper functions, the lower it gets.

(2) As the data setup is a prerequisite to be able to exercise your test, omissions in the data setup are most often revealed by the fact that the test cannot be successfully exercised.

Prebuilt, shared, or fresh fixture?

Chapter 5, Test Plan and Test Design, explained the differences between **prebuilt**, **shared**, and **fresh** fixtures and discussed which to use for what purpose. You might recall it was stated that "we will not use the **prebuilt fixture** pattern in any of our tests." There can, however, be a stringent practical reason to deviate from this: **test execution duration**. If each test is going to create all fixtures anew, this will increase the test execution duration substantially.

If the **fresh fixture** can be *promoted* to a shared fixture, as some data can easily be reused over a number of tests, do so. See also *Test example 4* in *Chapter 7, From Customer Wish to Test Automation – Next Level*. As you will see in the test example of this chapter, this promotion indeed can diminish the test execution duration.

Now, what about a **shared fixture** that is not only shared among tests in the same test codeunit, but also across multiple test codeunits? Recreating a shared fixture for each test codeunit run might also be a time-intensive one. So, also here, to reduce the test execution duration substantially, consider promoting it to become a prebuilt fixture.

> **Note**
> The main difference between promoting **shared fixture** to **prebuilt fixture** and using a pre-configured database like CRONUS is that our – promoted – shared fixture has been coded and, when promoting, the same applies to the prebuilt fixture. This way your tests stay in control of the data setup.

Examples of standard tests with complex data setup

Although we will work out an example for the **LookupValue** case, you might want to learn from the following standard test codeunits. For each of them, I have provided a short explanation on their data setup complexity and pointed out one specific test method, but I leave it to you to study the details:

- Codeunit 134202 - Document Approval - Users in Tests-User app

  This contains, among other tests, a nice collection of parameterized tests (see *Chapter 7*) with an extensive data setup for each test creating a complete document approvals workflow. Note the *teardown* at the end of each test and how standard code is triggered by calling an event publisher from the Approvals Mgmt. codeunit (1535).

  Example test method: ApproveRequestForPurchBlanketOrder

- Codeunit 134203 - Document Approval - Documents in Sales and Inventory Forecast Tests app

Another nice collection of tests with an extensive data setup for each test creating a complete document approvals workflow. Note how in many tests the setup is wrapped in very readable local helper methods and how, in some tests, standard code is triggered by calling an event publisher from the `Approvals Mgmt.` codeunit (1535).

Example test method: `SalesOrderWithPrepaymentApproval`

- `Codeunit 139540 - Sales Forecast Tests` in `Sales and Inventory Forecast Tests` app

A collection of well-structured tests with a great variety of data setups, both complex and simple. Note the low number of helper functions, the high number of UI handlers, and the usage of the test page `Trap` method, which allows you to catch the next invoked test page and assign it to the test page variable the `Trap` method is called from.

Example test method: `TestCreatePurchaseLineAction`

> **Note**
>
> In these examples, you might have noticed the different test-design patterns used and the difference in how the *test story is written*. Most likely you might have guessed that I am not a great fan of most of the coding. These could have been conceived with better readability. As such, I hope they do illustrate the value of my point that test code should be structured and formatted in the way I have been discussing so far. Nevertheless, they are still illustrative examples of complex data setup.
>
> Thanks to Microsoft's Georgios Papoutsakis for his suggestions for some of the examples here and below.

Verification

After the data setup, **verification** of your test is the second most time-consuming part of testing, in either simply manually executing or conceiving and implementing the test code. In *Chapter 5*, *Test Plan and Test Design*, I stressed the importance of the verification part of a test by stating that "a test without verification is no test at all." This illustrates that the verification – the checking and the testing – could be called the most essential part of a test. It also points to the significance of a well-thought-out set of checks to verify the outcome of a test. The complexity of this set lies more often than not in the task to find a sufficient set of checks that validates the complete output of the feature under test. A set that will stand the test of time. So, the complexity is not in the checks as such, but in the number of checks.

Contrary to an incomplete data setup, which has the exercise part as a safety net, an incomplete set of checks might not easily be unveiled. As mentioned in *Chapter 3, The Testability Framework*, a successful test is a test that does not fail. If your test has an insufficient set of checks that are all passing, your test passes, but it doesn't really verify the feature under test. Nothing is telling you that the set was insufficient and thus the test with it. With respect to the verification part – the THEN part(s) in your **ATDD** scenario – a great responsibility lies on the shoulders of the team. Any omission in this part when designing the tests might have far-reaching consequences: although running the tests gives us this cozy, green feeling, a big, red deficit might go unnoticed.

Using Library Assert

Regarding the verification part of your tests, I pointed to `Library Assert` (13002) in *Chapter 4, The Test Tools, Standard Tests, and Standard Test Libraries*. It contains a nice set of methods enabling you to check an *expected* value against an *actual* value. In case the check fails, it provides a standardized error message. These errors will clearly deviate from application and platform errors thrown in the *setup* and *exercise* part of your test. In a way, this lowers the complexity of failing verifications as the error message structure is easy to recognize. Reading the errors thrown by a `Library Assert` method, I very often can already make out why the test error occurred.

> **Note**
>
> My rough relative time estimation of getting verification done is 30%–40% of the work on tests.

Examples of standard tests with complex verification

Like with the previous examples, I have provided a short explanation of their verification complexity and pointed out one specific test method, but as before I leave it to you to study all the details:

- `Codeunit 135022 - Data Migration Facade Tests` in `Tests-Misc` app

  A collection of very detailed verifications. Note the repeat of the ATDD GIVEN-WHEN-THEN pattern in single tests, something I try to circumvent to let my test only have one WHEN part and thus, checking only one action. But as always, like the examples in the test codeunit, there are exceptions to the rule.

  Example test method: `TestCreateUpdateVendor`

- `Codeunit 137915 - SCM Assembly Posting` in `Tests-SCM` app

  A collection of not very well documented and difficult to read test methods. It however has a number of test methods with a very extended verification.

  Example test method: `VerifyPostedAssembly`

You might have noticed that in some test methods in the test codeunit, `Data Migration Facade Tests` (135022) THEN parts are placed before a WHEN, like in `TestCreateUpdateGLAccount`. It does make sense if it is somehow needed to verify the data setup. But looking at my own test code history and Microsoft's huge test collateral this can also be called an exception to the rule.

End-to-end code execution path

Since we discussed the complexity of **setup** and **verification** separately in the previous sections, it is logical to also have a look at the **exercise** part of tests – the WHEN in your **ATDD** scenario. This is the part where a test, given the setup, exercises the application under test to be able to verify if it behaves as expected. The *end-to-end code execution path* triggered by the exercise part can be quite a complex thing depending on the number of conditions it passes by. The longer the execution path under test, the higher chances are that this number is greater than 0, and the more complex it becomes to determine the specific path you want to test. Note that the exercise part itself will not be complex, it will in general just be a single action.

In a way, unit testing can be seen as a means of tackling *end-to-end code execution path* complexity. Testing the units is clearly less complex than testing the feature. But, as discussed in *Chapter 5*, *Test Plan and Test Design*, in the *And what about the unit and functional tests?* section, unit tests do not verify the feature as such. Even though unit tests might show that the – atomic – units are perfectly fine, it could be that the feature as a whole is not.

A typical example of a complex end-to-end code execution path in Business Central is the posting of a sales document – or a purchase document. This process seemingly has a zillion possible *linearly independent execution paths*, for example, posting different document types – quote, invoice, order, credit memo, blanket order, and return order – or checking the different mandatory fields. Extending this process can easily lead to a substantial number of tests, and the further into this process your extension intervenes, the more complex your test scenarios most likely will be.

> **Note**
>
> The complexity of an execution path can be measured by the so-called **cyclomatic complexity**. Cyclomatic complexity is expressed by the maximum number of linearly independent execution paths for the code under measurement. If the code contains no conditions the cyclomatic complexity is 1. When one condition is included and thus leads to two linearly independent execution paths, the cyclomatic complexity is 2. Want to read more? Go to `https://en.wikipedia.org/wiki/Cyclomatic_complexity`.
>
> For VS Code, a nice extension called AL Lint by Stefan Maroń is available that determines, among others, the cyclomatic complexity of methods.

Examples of standard tests with the complex end-to-end code execution path

Again, I have provided a short explanation of their complexity in the current context and pointed out one specific test method, and I leave again it to you to study all the details:

- `Codeunit 135404 - Sales Document Plan-based E2E` in `Tests-Permissions` app

  A bunch of very nicely ATDD structured tests with local helper method usage. Notice that (1) indeed the WHEN parts are simple, but the code it triggers is complex; and (2) again the use of `Trap` method

  Example test method:
  `TestCreatePurchaseOrderFromSalesOrderBusinessManager`

- `Codeunit 137067 - SCM Plan-Req. Wksht` in `Tests-SCM` app

  A well-structured test codeunit that clearly has been around for a while, as you can see from the majority of test methods using the **four-phase test** design pattern. Note that the readability could be improved by an extended usage of test helper methods.

  Example test method: `CalcPlanWithMPSForProductionForecastOrderItem`

Dependencies on other systems

Testing code that depends on other systems is a complexity on its own, not in the least as it is highly related to the availability of these systems at the time of executing your tests. As this is one of the topics of *Chapter 13, Testing Incoming and Outgoing Calls,* we will leave the discussion to that chapter, including the standard test examples.

Composing complex scenarios

Now that we have outlined what can make a scenario complex, it will be very practical to describe how we could *compose* complex scenarios enabling us to handle them efficiently and effectively. One after the other, the following are the hurdles to take to help you to get there:

1. Before a feature is being implemented, a **test plan** should be in place, detailed out in a **test design**, and reviewed by the team, as was discussed in *Chapter 5, Test Plan and Test Design*. Well-thought-out **ATDD** definitions will save a lot of work downstream.

2. Describe the **data setup** – the ATDD `GIVEN` parts – for your complex scenario, efficiently and effectively making use of a *common language* (see *Chapter 5*).

    Remove unneeded details and unveil missed details by team review.

    In this, make a clear split between this part – data setup – and the next – exercise. Determine what are the preparatory, data setup, steps needed to get the code under test exercised.

3. Determine which *linearly independent execution path* the **exercise** part – that is, the `WHEN` – of your complex scenario should follow. This defines, besides the data setup, additional input.

4. Describe all relevant **verification** criteria – the ATDD `THEN` parts – for your complex scenario, efficiently and effectively making use of a *common language* (see *Chapter 5*).

    And similar to *step 2*, remove unneeded criteria and unveil missed criteria using team reviews.

5. Now **implement** for your complex scenario the `GIVEN` parts **one after the other** and test in between.

    Tip: using the **ATDD.TestScriptor**, each `GIVEN` will get a helper method implemented with an error statement. This way you are helped to implement one after the other.

6. Now **implement** the `WHEN` part **and run the test** again.

    If it fails, it shows that either the `GIVEN` parts or additional input parameters to the `WHEN` part are not sufficient.

7. Now **implement** the `THEN` parts **one after the other** and test in between.

    Use *test the test* as introduced in *Chapter 6, From Customer Wish to Test Automation – the Basics*, to check the validity of each verification.

Walking through this list, you might wonder whether this only applies to complex scenarios. And you are fully right: it doesn't, it applies to any scenario. But… each stage in this list focuses on completing each stage well and starts with having a plan in place. And if there is *no plan*, there will be *no test*. Where you might manage to get away with not having a plan for less complex scenarios, you surely won't in case of complex scenarios.

In the following sub-sections, I will elaborate a bit more with respect to the steps above.

Finding common ground – steps 1, 2, 3, and 4

When getting the test plan and its related test design in place, it will be of great help to spend time in finding *common ground* in the GIVEN, WHEN, and THEN parts of the different scenarios:

- What ATDD tags do they have in common?
- What ATTD tags are very similar and could be molded into identical tags?

If possible, describe these tags exactly the same. This will allow you to reuse the resulting helper methods and speed up the implementation of your scenarios. The implementation of the first test might be very time consuming. I can easily spend hours on getting the first test worked out well. Once in place, however, I will find the second and next scenarios coming fast.

Deconstructing your scenario – steps 1, 2, 3, and 4

Put the time into deconstructing your (complex) scenario into separate and sequential tests. The big advantage of this is that your test will have a single purpose: checking one **exercise**. If it fails, it is only due to that single WHEN. If it succeeds, you know this part is working right. If you combine multiple WHEN parts in one test, a failing WHEN makes the whole test fail. You might, however, have noticed that in various standard test examples referenced in the previous section, Microsoft implemented multiple GIVEN-WHEN-THEN sequences in single tests. Indeed, this works fine, but as pointed out above, this should be the exception to the rule. I would rather split up such a test into a consecutive sequence of dependent tests that need to be run as a whole and in this sequence. This way you will get a clear report on what parts succeeded and what did not. If you disagree with me, then at least exercise my advice while developing the test(s) and once ready and OK, aggregate the sequence of dependent tests into one test.

> **Note**
> Creating, maintaining, and executing a consecutive sequence of dependent tests has a couple of challenges as there is now a formal way to link them together and enforce their execution to be always a complete run of these tests. To ensure this, however, as much as possible place such a chain of tests in a dedicated test codeunit and name each test in such a way that it is clear that each of them is a step in a chain of tests.

Using flowcharts – steps 1 and 3

To be able to determine the *linearly independent execution paths* that your scenarios are going to cover, flowcharts are a great help in this. If your feature is started from scratch this allows you and your team to determine, discuss, and review all different execution paths. This also enables you to set up a fairly complete test plan. If you're hooking into an existing process, for example, warehouse shipment posting, you might do without. In this case, you are able to run the process, while debugging, and get a fair enough understanding of the different execution paths. But I guess you do get the point that flowcharts will be of value in this case too.

> **Note**
> In this GitHub repo, `https://github.com/fluxxus-nl/Test-Automation-Examples`, you will find three examples of ATDD-based implementations that also made use of flowcharts.

Using Code Coverage – steps 1 and 3

In *Chapter 9*, *How to Integrate Test Automation in Daily Development Practice*, I showed how the **Code Coverage** tool can be of help to measure what code is being covered by the tests you have implemented. As said, it is, however, not an unambiguous measure. The value of the code coverage measured lies not so much in the code covered, but in the code not yet covered. As such, the Code Coverage tool can help you to uncover *linearly independent execution paths* not yet in your test plan.

Using standard helper functions – steps 5, 6, and 7

Once the test design for your (complex) scenario is agreed upon to be sufficiently complete, it's time to go and implement it. Given you have found the *common ground* for the GIVEN, WHEN, and THEN parts of the different scenarios as discussed above, you will be able to implement each tag efficiently and effectively. While doing this make as much use as possible of standard helper functions provided by Microsoft in their vast collection of tests. This will take a lot of work off your hands. No need to implement – and maintain! – a whole bunch of creator methods for use in GIVEN parts. Likewise, no need to code generic process triggering, like document posting or MRP runs, to be used in the WHEN part.

As you have seen, I gratefully used a number of standard creator functions in most of the aforementioned test examples. In the *Test example 10* section, I will extend this by using a number of process triggering helper functions.

Creating reusable parts

One of the reasons software development and automation has such a prominent place in today's world is its repeatability. Recurring tasks can be executed in exactly the same way repeatedly. But it's not only this kind of repeatability that's responsible for automation's prominent place. It's also the ability to create reusable parts. Now, you might recall that in practicing TDD you, at first, do not focus on generalization, that is, creating reusable parts, but let this emerge as part of the refactoring. And that is fully true. So, with TDD we first focus on getting the test(s) implemented. This, however, does not prohibit us from designing our test in such a way that we reveal and make use of *common ground* as discussed in *Finding common ground – steps 1, 2, 3, and 4*.

Test example 10 – how to construct an extensive scenario

To give an illustration of the above, we elaborate on another part of our customer wish.

> **Note**
>
> In this example, we focus on the specifics discussed above, not on the whole implementation – test and app code – of the scenarios. As such, our discussion will not follow the order of the TDD **red**-**green**-**refactor** mantra and will only include code if it is relevant for the context. The completed test and app code can be found in the GitHub repo.

Customer wish

In the business logic description of our customer wish it is mentioned that:

When creating a warehouse shipment from a sales order, the Lookup Value Code field should be inherited from the Sales Header to the Warehouse Shipment Line.

This is expressed in the following two scenarios:

```
[FEATURE] LookupValue Warehouse Shipment
[SCENARIO #0030] Create warehouse shipment from sales order with
                 lookup value
[GIVEN] Lookup value
[GIVEN] Location with require shipment
[GIVEN] Warehouse employee for current user
[WHEN] Create warehouse shipment from released sales order with
       lookup value and with line with require shipment location
```

```
[THEN] Warehouse shipment line has lookup value code field
       populated
```

```
[SCENARIO #0031] Get sales order with lookup value on warehouse
                 shipment
[GIVEN] Lookup value
[GIVEN] Location with require shipment
[GIVEN] Warehouse employee for current user
[GIVEN] Released sales order with lookup value and with line
        with require shipment location
[GIVEN] Warehouse shipment without lines
[WHEN] Get sales order with lookup value on warehouse shipment
[THEN] Warehouse shipment line has lookup value code field
       populated
```

These, however, imply that the `Lookup Value Code` field already exists in both the `Warehouse Shipment Line` and `Posted Whse. Shipment Line` tables, which is defined by the following three fundamental scenarios on our list, which we have skipped so far:

```
[SCENARIO #0015] Assign lookup value to warehouse shipment line
[GIVEN] Lookup value
[GIVEN] Location with require shipment
[GIVEN] Warehouse employee for current user
[GIVEN] Warehouse shipment from released sales order with line
        with require shipment location
[WHEN] Set lookup value on warehouse shipment line
[THEN] Warehouse shipment line has lookup value code field
       populated
```

```
[SCENARIO #0016] Assign non-existing lookup value on warehouse
                 shipment line
[GIVEN] Non-existing lookup value
[GIVEN] Warehouse shipment line record variable
[WHEN] Set non-existing lookup value to warehouse shipment line
[THEN] Non existing lookup value error was thrown
```

```
[SCENARIO #0017] Assign lookup value to warehouse shipment line
                 on warehouse shipment document page
```

```
[GIVEN]  Lookup value
[GIVEN]  Location with require shipment
[GIVEN]  Warehouse employee for current user
[GIVEN]  Warehouse shipment from released sales order with line
         with require shipment location
[WHEN]   Set lookup value on warehouse shipment line on warehouse
         shipment document page
[THEN]   Warehouse shipment line has lookup value code field
         populated
```

As the scenarios are related, it won't be a surprise to find congruous parts. It's a message happily notifying us that we will be able to construct *reusable* parts and save time when working out all five scenarios.

These are the obvious ones:

```
[GIVEN]  Lookup value
[GIVEN]  Location with require shipment
[GIVEN]  Warehouse employee for current user
```

And you can often find other hints inline. Compare the following taken from four of the five scenarios:

```
[SCENARIO #0015]
[GIVEN]  Warehouse shipment from released sales order with line
         with require shipment location


[SCENARIO #0017]
[GIVEN]  Warehouse shipment from released sales order with line
         with require shipment location


[SCENARIO #0030]
[WHEN]   Create warehouse shipment from released sales order with
         lookup value and with line with require shipment location


[SCENARIO #0031]
[GIVEN]  Released sales order with lookup value and with line
         with require shipment location
```

This tells us that, in all these cases, we will need a released sales order with a lookup value.

And another hint leading to another reusable part: compare the following taken from the same four scenarios. Do you see what I mean?

[SCENARIO #0015]

[THEN] Warehouse shipment line has lookup value code field
       populated

[SCENARIO #0017]

[THEN] Warehouse shipment line has lookup value code field
       populated

[SCENARIO #0030]

[THEN] Warehouse shipment line has lookup value code field
       populated

[SCENARIO #0031]

[THEN] Warehouse shipment line has lookup value code field
       populated

Application code

On account of scenarios #0015, #00016, and #0017, the Warehouse Shipment Line and Posted Whse. Shipment Line tables and their related pages should be extended with the **Lookup Value Code** field. This is similar to what has been done in the test examples of *Chapter 6, From Customer Wish to Test Automation – the Basics*, and *Chapter 9, How to Integrate Test Automation in Daily Development Practice*. To study the details, refer to the code in the GitHub repo.

To enable scenarios #0030 and #0031, the standard application has been extended with the following code to ensure that the value of the Lookup Value Code field on a sales document is copied to the Lookup Value Code field in the warehouse shipment line:

```
codeunit 50002 "WhseCreateSourceDocumentEvent"
{
    [EventSubscriber(ObjectType::Codeunit,
      Codeunit::"Whse.-Create Source Document",
      'OnBeforeCreateShptLineFromSalesLine', '', false, false)]
    local procedure OnBeforeCreateShptLineFromSalesLineEvent(
      var WarehouseShipmentLine:
        Record "Warehouse Shipment Line";
```

```
    WarehouseShipmentHeader:
      Record "Warehouse Shipment Header";
    SalesLine: Record "Sales Line";
    SalesHeader: Record "Sales Header")
  begin
    WarehouseShipmentLine."Lookup Value Code" :=
      SalesHeader."Lookup Value Code";
  end;
}
```

Test code

With the code on GitHub, we leave it to you to have a look at the implementation of the test code for the four overlapping scenarios #0015, #0017, #0030, and #0031. This code can be found in the test codeunit LookupValue Warehouse Shipment (81003 – see object file LookupValueWarehouseShipment.Codeunit.al).

Here we'll have closer look at a number of details of scenario #0030, wherein both the GIVEN and THEN parts are shared with the other three scenarios. In this section, we show you how we create and use our own reusable parts and apply some *reusable parts* – that is, helper functions – from the standard libraries.

Initialize

In the Initialize function, being the first reusable component that is shared, part of the data setup is being handled:

```
[GIVEN] Lookup value
[GIVEN] Location with require shipment
[GIVEN] Warehouse employee for current user
```

In code:

```
local procedure Initialize()
var
  WarehouseEmployee: Record "Warehouse Employee";
begin
  if isInitialized then
    exit;

  //[GIVEN] Lookup value
  LookupValueCode := LibraryLookupValue.CreateLookupValueCode();
```

```
//[GIVEN] Location with require shipment
LibraryWarehouse.CreateLocationWMS(
  DefaultLocation, false, false, false, false, true);
//[GIVEN] Warehouse employee for current user
LibraryWarehouse.CreateWarehouseEmployee(
  WarehouseEmployee, DefaultLocation."Code", false);

isInitialized := true;
Commit();
end;
```

The structure of `Initialize` by now should look familiar, including the usage of the global Boolean variable `isInitialized`, the global `Code[10]` variable `LookupValueCode`, and the local helper function, `CreateLookupValueCode`. I reckon, as shown in test example 4, you're also familiar with the fact that `CreateLookupValueCode` can be part of the lazy setup.

The creation of a location (with *require shipment*) and a warehouse employee can also be embedded in `Initialize` as these can easily be shared between the four scenarios. For this, `DefaultLocation` is set up as a global record variable (based on the `Location` table). The warehouse employee does not need to be stored as a variable as it will be retrieved from the database.

As you can see, we made use of two helper functions through a library codeunit variable `LibraryWarehouse` based on the standard codeunit called `Library - Warehouse`. Using the simple and quick file search method as mentioned in *Chapter 4, The Test Tools, Standard Tests, and Standard Test Libraries*, I hunted for a helper function to create a location, search string `CreateLocation`, and one to create a warehouse employee, search string `CreateWarehouseEmployee`. For an extensive search example, see the *Finding the standard helper functions* section.

Running the test functions for the scenarios `#0015`, `#0017`, `#0030`, and `#0031` shows that making use of the `Initialize` function is not only a matter of creating code that is easier to maintain and understand, but also much faster, which, as I pledged earlier in the book, is a must for automated tests. The following image summarizes the average duration of each scenario:

Scenario (execution order)	Average Duration with fresh setup (ms)	Average Duration with lazy setup (ms)	Delta %
#0015	177,1	174,8	-1%
#0017	209,1	151,3	-38%
#0030	144,7	91,8	-58%
#0031	145,9	98,0	-49%
Total	169	129	-31%

Figure 11.1 – Average duration of each scenario of test example 10

Having run all four tests 10 times individually (and thus triggering `Initialize` as if it were a fresh setup), and also having run the four tests in one go 10 times (now triggering `Initialize` to get a shared setup), shows that the latter is more than 30% faster.

Notes

(1) In both cases, fresh and lazy setup, scenario #0015 is just as fast because it always makes `Initialize` run fully.

(2) Scenario #0016 was left out of this test as it does not contain a call to the `Initialize` method.

VerifyLookupValueOnWarehouseShipmentLine

`VerifyLookupValueOnWarehouseShipmentLine` implements the reusable THEN part pointed out above:

```
[THEN]  Warehouse shipment line has lookup value code field
         populated
```

So, with `VerifyLookupValueOnWarehouseShipmentLine` the second reusable part is found. It closely resembles the various `VerifyLookupValueOn` helper functions in the previous examples. Hence, with practicing the Business Central developer's virtue, it's a quick task to code `VerifyLookupValueOnWarehouseShipmentLine`: copy, paste, and adjust. We leave that to you to either perform this virtue or reference the ready code in the GitHub repo.

CreateWarehouseShipmentFromSalesOrder

If you have experience with the warehouse shipping feature of Business Central, you know that a series of steps have to be performed to get a warehouse shipment created. It's not as single-fold an operation as creating a purchase invoice. To get the *execution path* well defined, drawing the flow of this process is a great help. The schema in *Figure 11.2* displays the tasks that need to be performed in the last GIVEN of the scenarios #0015 and #0017, and in the WHEN part of scenario #0030:

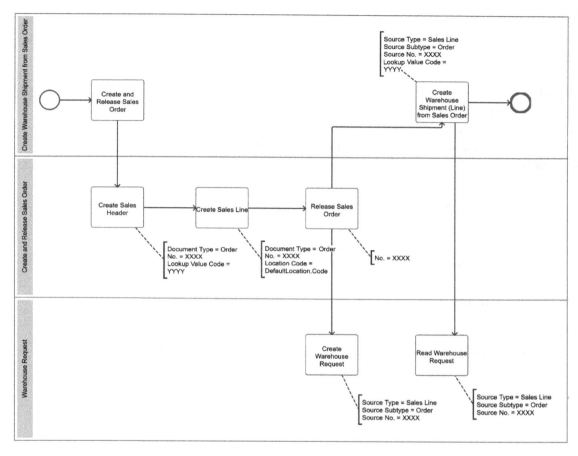

Figure 11.2 – Flowchart for the process of creating a warehouse shipment from a sales order

Based on this schema, we created well-readable, reusable, minimalistic functions:

- `CreateWarehouseShipmentFromSalesOrder` is being used by scenarios #0015, #0017, and #0030.

- `CreateAndReleaseSalesOrder` (see the middle swim lane) is directly being used by scenario #0031 and indirectly by #0015, #0017, and #0030.

Being a standard Business Central process, it will not come as a surprise that we did not only create our own reusable parts, `CreateWarehouseShipmentFromSalesOrder` and `CreateAndReleaseSalesOrder`, but with them, we also made use of a number of standard helper functions from the Microsoft test libraries.

Finding the standard helper functions

Let's use this as an illustrative example of how I go about finding relevant helper functions in the Microsoft test libraries.

Create a sales document with an item and location

The first two actions in the second swim lane in *Figure 11.2* tell me that we need a sales order with a line containing an item and a location. Knowing Business Central, and of course also somewhat of the test libraries, I am quite sure there exists a helper function that can get this done for me. Or at least the main part. So, I will access a test (library) source folder with VS Code and do a file search for:

```
^\s*procedure CreateSalesDocumentWith
```

> **Note**
> I am using a regular expression to search only for non-local procedures.

Resulting in three hits:

- `ERMCurrencyFactor.Codeunit.al`

  `procedure CreateSalesDocumentWithoutFCYExchangeRate()`
- `ERMVATToolHelper.Codeunit.al`

  `procedure CreateSalesDocumentWithRef()`
- `LibrarySales.Codeunit.al`

  `procedure CreateSalesDocumentWithItem()`

The top two functions are in test codeunits and as such are no candidates as these are test functions and of course not meant to be referenced. This is leaving us only the third result from `Library - Sales`. This is what I used.

Release a sales document

The third action in the second swim lane tells me we need a helper function allowing us to release a sales document.

My search string:

```
^\s*procedure ReleaseSalesDoc
```

Resulting in only one hit:

`LibrarySales.Codeunit.al`

`procedure ReleaseSalesDocument()`

Sounds exactly like what we need.

Create a warehouse shipment from sales order

As per the action in the first swim lane, we need a helper function that can create a warehouse shipment for our sales order.

My search string:

`^\s*procedure Create(Warehouse|Whse)ShipmentFrom`

Resulting in seven hits:

- `ServiceOrderRelease.Codeunit.al`

  `procedure CreateWarehouseShipmentFromGetSourceDocument()`
- `SCMPrepaymentOrders.Codeunit.al`

  `procedure CreateWarehouseShipmentFromSalesOrderWithPostPrepayment Invoice()`
- `SCMWarehouseShipping.Codeunit.al`

  `procedure CreateWarehouseShipmentFromSalesOrderWithMultipleLocation()`
- `LibraryWarehouse.Codeunit.al`

  `procedure CreateWhseShipmentFromPurchaseReturnOrder()`

  `procedure CreateWhseShipmentFromServiceOrder()`

  `procedure CreateWhseShipmentFromSO()`

  `procedure CreateWhseShipmentFromTO()`

The first three hits are all in test codeunits and, thus, like explained in *Create a sales document with an item and location* section, are no candidates. The latter four, however, all reside in a library codeunit and the one that fits us is the third one: `CreateWhseShipmentFromSO`.

Test execution

We have found all the helper functions we need. We have the app and test code ready. Let's do a final run: run, runn, runnn, runnnnn… **grrrrrrreeeeeeen**!

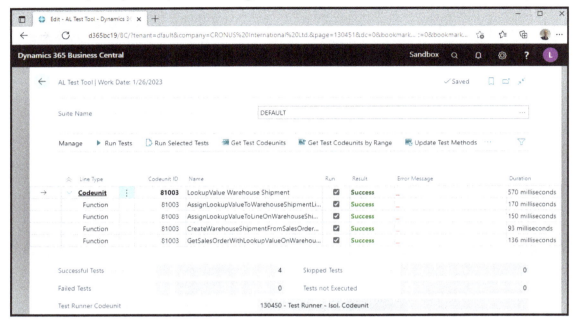

Figure 11.3 – Test example 10 having been run: GREEN

Summary

In this chapter, you learned what things can make a scenario complex and how to go about handling them efficiently and effectively, finding *common ground*, *deconstructing* it, and using *flowcharts*, *code coverage*, and *reusable parts*, the latter being either standard helper functions or ones you create yourself.

In the next chapter, *Chapter 12, Writing Testable Code*, we will discuss how to write code that is better and more testable with test automation.

12
Writing Testable Code

So far, in this book, our focus has been on writing test code. The **LookupValue** extension, our example case, came with application code mainly ready. As such we did not talk about how to get your application code better suited to being tested. We did not ask ourselves: what does it mean to write testable code?

In this chapter, we'll find an answer to this query based on the following questions:

- What is testable code?

- How to make your code testable?

As we have done in the previous chapters, we will go through a test example, *Test example 11 – how to unit test event subscribers' logic*, where we'll work out one of the techniques discussed.

Technical requirements

Like in the previous chapter, we will refer to the **LookupValue** extension. Its code is to be found on GitHub: `https://github.com/PacktPublishing/Automated-Testing-in-Microsoft-Dynamics-365-Business-Central-Second-Edition`. In this chapter, we will continue with the code from *Chapter 11, How to Construct Complex Scenarios*: `https://github.com/PacktPublishing/Automated-Testing-in-Microsoft-Dynamics-365-Business-Central-Second-Edition/tree/main/Chapter 11 (LookupValue Extension)`.

The final code of this chapter can be found at `https://github.com/PacktPublishing/ Automated-Testing-in-Microsoft-Dynamics-365-Business-Central- Second-Edition/tree/main/Chapter 12` (LookupValue Extension).

Details on how to use this repository and how to set up VS Code and the Business Central environment are discussed in *Appendix, Getting Up and Running with Business Central, VS Code, and the GitHub Project*.

What is testable code?

Almost two decades ago, in 2002, Microsoft acquired the Danish company Navision A/S, at that time a relatively recent merger of Navision Software A/S and Damgaard A/S. You could say that with this acquisition my testing journey started, because – after I rounded off my previous role as a technical trainer at the Dutch Navision Academy – I joined the former Dynamics NAV **Global Development Localization (GDL)** team at Microsoft as an application tester and documenter. Early 2005, I was taking part in two Microsoft internal, so-called, *Engineering Excellence* courses; one for new Software Engineers and one for Test Leads. It was a great experience, not in the least due to the enthusiasm and great experience of our trainer Tracy Monteith (`https://www.linkedin.com/in/tracy-monteith-021868115/`). There are many things I owe to him, but one of the most prominent things that has stuck with me ever since is the notion of **testable code**. I can still recall a certain astonishment when Tracy stated that code should be *testable*. It sounded to me like he was saying in all earnestness that we needed to breathe today. My mind ran like crazy asking myself:

Isn't code always testable?

And if it wasn't what was I doing as a tester?

And it continued:

If I can test code, it is testable, right?

Logical questions from my context as a tester, a manual tester at that time. Indeed, I can test the code, and thus call it testable, but clearly that was not what Tracy was talking about.

Testable code is code that can be tested efficiently and effectively using ... code, that is, **test automation**.

Testable code allows you to have full control over checking the behavior from your test code. If, for example, you want to test whether your code handles the interaction with an external system right, you don't want to be bothered by the interaction itself as that's not what you are testing. Your tests should not rely on the availability of the external system. Rather, you want to mimic it's there, so, your test code can control whether it returns a requested response, or, just as important, an unexpected response, and see if the software under test handles it as expected.

Knowing Business Central as a platform and an application, I could imagine your mind running and wondering: how about Business Central and testable code? Do they go together? Indeed, there are various architectural attributes of the platform and the application that limit and complicate our ability to write fully testable code for Business Central, like:

- A major part of the standard code is highly coupled and often our features need to hook into these.

- Application and database are highly intertwined, there is no standard way to abstract the data layer, to enable us to mimic data retrieval from our database.

- A platform that for so long, and to some extent still, did not provide advanced coding techniques, like object-oriented programming, to enable writing testable code.

Notwithstanding these arguments, we'll discuss next how you can make your, existing and new, code testable.

> **Note**
>
> In *Chapter 13, Testing Incoming and Outgoing Calls*, we will have a look at how to test externally accessible components like APIs.

How to make your code testable?

Having full control of the code under test, in essence, means that this code has no direct dependency on other pieces of code in your application and outside it. In other words: your code is broken up into loosely coupled units. Either they are fully independent or, if a dependency applies, your code allows you to control the dependency. Microsoft has been highly investing, and still is, in getting the application loosely coupled and better testable, like they have done when splitting the system and base application. They also continue doing this with the base application itself where there is still a vast number of battles to be won like in the case of the sales posting routine. Although it has been refactored significantly, there is a still massive blob of code that executes 1,000 tasks in the process of posting a sales order. We can hardly access just parts of the code, it is only possible to execute the whole thing. Imagine if the code is executed as a sequence of loosely coupled individual function calls, each unit could then be tested on its own. Isn't this what is called **unit testing**? Thus, to its full extent, enabling unit testing is about writing testable code.

Coding techniques

Now let's have a look at the following *coding techniques* that will help to make your code testable, starting with "low-hanging fruit" and moving to a more advanced level:

1. Extract implementation of business logic into testable methods.

2. Reduce the number of code execution paths in each method.

3. Replacement of dependencies.

Extract implementation of business logic into testable methods

Historically, implementing new business logic in Business Central was very often done by coding in the various object triggers the platform has been providing for ages. Triggers like OnInsert, OnModify, OnDelete, and OnRename on tables, OnValidate and OnLookup on fields, and OnOpenPage and OnAction on pages and page actions. Now, getting this new business logic exercised as part of your testing means that you need to get the specific trigger invoked, which most likely can only be achieved by triggering the object the code resides on. To enable, for example, the execution of a process coded on a page action, you need to open the page and invoke the action, as a minimum. From a manual testing practice, this is no issue as this is literally the only way you will get it started. By now, I gather, you do understand that from a test automation perspective this is not the most efficient way to get it done. You rather want to extract the implementation of the business logic into a **testable method**. At the same time, you might recognize that this extraction does not only serve the testability of your code, but also reusability and readability. The next code segment, taken from the Customer table, shows a nice, illustrative, and comprehensible example in the OnValidate trigger of the Country/Region Code field:

```
field(35; "Country/Region Code"; Code[10])
{
  Caption = 'Country/Region Code';
  TableRelation = "Country/Region";

  trigger OnValidate()
  begin
    PostCode.CheckClearPostCodeCityCounty(
      City, "Post Code", County,
      "Country/Region Code",xRec."Country/Region Code");

    if "Country/Region Code" <>
      xRec."Country/Region Code"
    then
```

```
        VATRegistrationValidation();
    end;
}
```

Two business rules have been implemented in two isolated methods – highlighted in the code segment above – with a clear intent, *basically* making them independent testable units:

- `CheckClearPostCodeCityCounty`, to check `Post Code`, `City`, and `County` fields and, if applicable, clear them.

- `VATRegistrationValidation`, to validate the `VAT Registration No.` field in case the `Country/Region Code` field has been changed.

Note that I put stress on *basically* as there is more to be said about the methods themselves, but I'll leave that for later.

Using event subscribers as a means to isolate testable units

In Dynamics NAV 2016, Microsoft introduced events, allowing you to extend existing business logic in a highly decoupled way. With the transition into full AL development, this has become the only way to do this, like, for example, the Business Central standard application. Given an event publisher, you extend the existing business logic by linking an event subscriber to it and triggering your additional business rules from this event subscriber, like we did in the **LookupValue** extension too. Such an event subscriber could be seen as a testable unit to verify these additional business rules. Very often, however, an event subscriber will not comply with the **single-responsibility principle (SRP)** defined as the first principle – the **S** – of the **SOLID** design principles, meaning that it services more than one goal; or in other words: multiple business rules are implemented in this subscriber. It is therefore best practice to not implement the business logic directly into the subscriber but rather in public methods that address each business rule. Each method then has a clear intent, complies with **SRP**, and is, as such, a testable unit. If needed it could be "extended" by overloading. Note that the latter you cannot do with event subscribers as they are bound to their contract with the publisher.

In *Test example 11 – how to unit test event subscribers' logic*, placed at the end of this chapter, you can see how you can go about doing this.

> **Note**
>
> Trying to find a balance between focusing on writing testable code as such and the various coding principles behind that, I decided not to go too deep into **SOLID**. If you, however, want to learn more about it, two reading suggestions: (1) `https://en.wikipedia.org/wiki/SOLID` and (2) `https://medium.com/feedzaitech/writing-testable-code-b3201d4538eb`.

Reduce the number of code execution paths in each method

Writing your application code in testable units is one thing, potentially making the application simpler to test. Another is to make those units themselves simpler to test. One aspect of this is that it should not take too many tests to verify the validity of the unit. To achieve this, reduce the number of code execution paths in your method. The more code execution paths there are, the more scenarios you have to check. This potentially can lead to a so-called **combinatorial explosion** (`https://en.wikipedia.org/wiki/Combinatorial_explosion`). Too many scenarios to – reasonably – cover all the code. Of course, if the number of code execution paths is already at a minimum there is nothing to reduce, but often we're not aware whether that's the case.

Cyclomatic complexity is a very useful measure to help you reduce the number of code execution paths. See my note in *Chapter 11, How to Construct Complex Scenarios*, for a short description of what it entails. You also might want to read this nice article by the famous David Bernstein: `https://dzone.com/articles/making-code-testable`.

> **Note**
> Having a lot of execution paths is sometimes a smell of a broken **SRP**. What potentially led to having so many execution paths, is the fact that your code has multiple responsibilities, and the combination of those responsibilities is what explodes.

Replacement of dependencies

Another aspect of making your units simple – or even feasible – to test, is the replacement of dependencies on other components. For this, we have two options, a simple and an advanced one:

- **Simple** – Remove the dependency by handing over specific values when calling your unit. You might want to call this *remove dependency*.

- **Advanced** – Replace the dependency when calling your unit, which is also known as **dependency injection** (**DI**).

Remove dependency

In general, the result – or output – of your unit depends on the input it gets. This input can be explicitly defined by the signature of the unit – its parameters – but is often also defined by implicit references to things like a default setup. This is a very common pattern in Business Central code. This is an example of why Business Central in general is not well, that is efficiently, testable with test automation, as, before you can test your unit, it could be necessary to update data in various tables.

Example – Remove dependency

To make it more concrete with a simple example, let's study one of the **testable methods**, shortly discussed in *Extract implementation of business logic into testable methods*, somewhat closer. Looking at the statement calling `CheckClearPostCodeCityCounty`, it seems to be a method with a clearly defined contract:

```
PostCode.CheckClearPostCodeCityCounty(
    City, "Post Code", County,
    "Country/Region Code", xRec."Country/Region Code");
```

Provided a `City`, `Post Code`, `County`, and current and previous `Country/Region Code`, it will do its job. Examining the method itself does support this conclusion. Have a look yourself:

```
procedure CheckClearPostCodeCityCounty(
    var CityTxt: Text; var PostCode: Code[20];
    var CountyTxt: Text;
    var CountryCode: Code[10]; xCountryCode: Code[10])
var
    GLSetup: Record "General Ledger Setup";
    IsHandled: Boolean;
begin
    IsHandled := false;
    OnBeforeCheckClearPostCodeCityCounty(
        CityTxt, PostCode, CountyTxt,
        CountryCode, xCountryCode, IsHandled);
    if IsHandled then
        exit;

    GLSetup.Get();
    if GLSetup."Req.Country/Reg. Code in Addr." then
        exit;

    if (xCountryCode = CountryCode) or (xCountryCode = '') then
        exit;

    PostCode := '';
    CityTxt := '';
    CountyTxt := '';
end;
```

At first glance, `CheckClearPostCodeCityCounty` might appear to operate fully upon the data provided by its contract: `CityTxt`, `PostCode`, `CountyTxt`, `CountryCode`, and `xCountryCode`. But quickly after – or maybe even during – this first glance, you might notice that there is more data used: the `General Ledger Setup`, or more specifically its `Req.Country/Reg. Code in Addr.` field. Our testable method clearly has a dependency on this making it less testable. If we are we to write code to test the method `CheckClearPostCodeCityCounty`, this code first has to set the `Req.Country/Reg. Code in Addr.` field on the `General Ledger Setup`, get it updated in the database, and call `CheckClearPostCodeCityCounty`, which then retrieves the `General Ledger Setup` through the `GLSetup.Get()` statement.

From a testability perspective, this is quite a costly roundtrip that could be evaded. In this case, we have two options:

1. Make `General Ledger Setup` part of the contract.
2. Make `Req.Country/Reg. Code in Addr.` part of the contract.

As the latter is much simpler –, less costly – handing over a Boolean value instead of a whole record, and more to the point, the signature of `CheckClearPostCodeCityCounty` would then become:

```
procedure CheckClearPostCodeCityCounty(
   var CityTxt: Text; var PostCode: Code[20];
   var CountyTxt: Text;
   var CountryCode: Code[10]; xCountryCode: Code[10];
   ReqCountryCodeInAddr: Boolean)
```

We could then, of course, do away with the `GLSetup` variable and its `Get` statement. The upsides are that we now have better control over calling `CheckClearPostCodeCityCounty` and a faster execution when testing. The downsides – as each upside goes with a downside – will be that when calling `CheckClearPostCodeCityCounty` the general ledger setup probably needs to be retrieved and each call to this method needs to be refactored, being at approx. 30, in the standard Business Central application.

> **Note**
>
> The `General Ledger Setup` is an implementation of the Singleton pattern. As such it's a table containing a single record that in general resides in the cache of the Business Central service tier as it is very often referenced and hardly ever changed. When your code updates it, it will be sent to the databases, but it's still readily available to your code from the service tier cache.

Dependency injection (DI)

With **DI** you move the instantiation of a dependency – referred to as *service* – outside of the unit – referred to as *client*. This way you allow the dependency to be set at the moment you are going to call the client. As such **DI** gives you control over and flexibility in using the right services. This is exactly one of the major reasons to use dependency injection in test automation: the need to control the output of the component your unit is depending on.

If the *service* is, for example, an external system that might not be available at the time you're running your tests, the outcome of your test run will not be a success. The problem is that this failure shows nothing about the execution of the code under test, so it will most likely be of little use as a valid test result. Within the context of the coded test, the right service might be a so-called *mock*, that is, a component that mimics a real service and only executes things relevant for your tests. Thus, with the call of the unit – the *client* – from your test code, you inject a mock that will act as the *service*.

A simple, straightforward AL usage of this principle is to make the call upon a certain process from your unit based on a codeunit or report ID, for example, the `SendToPosting` method on the `Sales Header` (36) table:

```
procedure SendToPosting(PostingCodeunitID: Integer)
  IsSuccess: Boolean
var
  ErrorContextElement:
    Codeunit "Error Context Element";
  ErrorMessageMgt:
    Codeunit "Error Message Management";
  ErrorMessageHandler:
    Codeunit "Error Message Handler";
begin
  if not IsApprovedForPosting then
    exit;

  Commit();
  ErrorMessageMgt.Activate(ErrorMessageHandler);
  ErrorMessageMgt.PushContext(
    ErrorContextElement, RecordId, 0, '');
  IsSuccess := Codeunit.Run(PostingCodeunitID, Rec);
  if not IsSuccess then
    ErrorMessageHandler.ShowErrors;
end;
```

Testing the `SendToPosting` method as such does not need us to provide the ID of a real posting codeunit. It suffices to provide the ID of a *mock* codeunit that only needs to return true, that is, run without an error, or false, having produced an error that will be shown through the `ErrorMessageHandler` and nothing more. A nice tradeoff of **DI** usage in tests is the mock having far less logic to execute, making the test faster than running the "real" service even, if possible.

A better, but maybe a bit more sophisticated, way is to define an interface that you add as a parameter to your unit and, when calling the unit, hand over a specific implementation of the interface.

Note

Dependency injection (DI) is a form of the **dependency inversion principle**, the last principle – the **D** – of the **SOLID** design principles.

If you want to read more on **DI** go to `https://en.wikipedia.org/wiki/Dependency_injection`. Follow various embedded links for more details.

Example – dependency injection

Let's also make this concrete using the other **testable method**, discussed in *Extract implementation of business logic into testable methods*: `VATRegistrationValidation`.

Contrary to `CheckClearPostCodeCityCounty`, which we discussed just now, `VATRegistrationValidation` seems to have no contract. That is: it has no arguments that allow us to hand over anything when calling it from our test code. Studying its implementation, the first things that might strike your eye – in the context of dependencies – are the local variables. Especially the record and codeunit variables, highlighted in the next code snippet:

```
procedure VATRegistrationValidation()
var
  VATRegistrationNoFormat:
    Record "VAT Registration No. Format";
  VATRegistrationLog: Record "VAT Registration Log";
  VATRegNoSrvConfig: Record "VAT Reg. No. Srv Config";
  VATRegistrationLogMgt:
    Codeunit «VAT Registration Log Mgt.»;
  ResultRecordRef: RecordRef;
  ApplicableCountryCode: Code[10];
  IsHandled: Boolean;
  LogNotVerified: Boolean;
```

These clearly define four dependencies and do raise a major question that should be asked when striving to write testable code:

Can't these dependencies be injected?

Or maybe better: shouldn't one or more of these dependencies be injected? For this let's first have a closer look at the broader context.

The `VAT Reg. No. Srv Config` table – accessed through the **EU VAT Registration No. Validation Service Setup** page displayed in *Figure 12.1* – enables you to set up the details of an online service for:

… validating VAT identification numbers of economic operators registered in the European Union for cross-border transactions on goods and services.

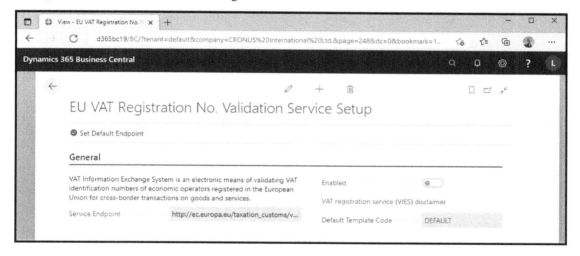

Figure 12.1 – EU VAT Registration No. Validation Service Setup

When the service is enabled, the following code will be triggered when calling `VATRegistrationValidation`:

```
if VATRegNoSrvConfig.VATRegNoSrvIsEnabled then begin
  LogNotVerified := false;
  VATRegistrationLogMgt.ValidateVATRegNoWithVIES(
  ResultRecordRef, Rec, "No.",
  VATRegistrationLog.
    "Account Type"::Customer.AsInteger(),
  ApplicableCountryCode);
  ResultRecordRef.SetTable(Rec);
end;
```

Even though its contract does not reveal it, the highlighted method `ValidateVATRegNoWithVIES` will – indirectly – make a call on the service defined in the **Service Endpoint** on the **VAT Registration No. Validation Service Setup** page. Now, testing the previous code will only succeed when that external service indeed is reachable. If it's not available, the test will fail as is also displayed in *Figure 12.2* when manually creating a new EU customer and needing to provide the VAT registration number:

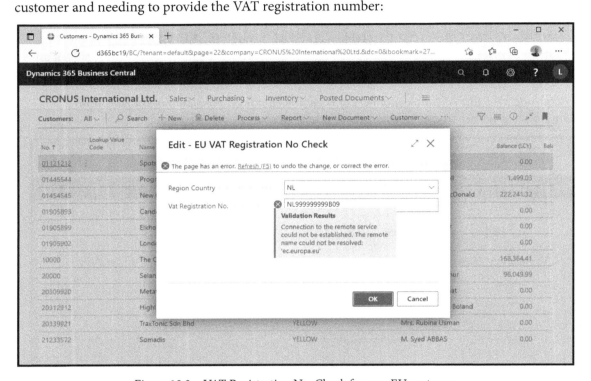

Figure 12.2 – VAT Registration No. Check for new EU customer

This manual test is a failed test, but its result is meaningless because this particular failure has nothing to do with the code under test. Thus, the success of a coded test checking `VATRegistrationValidation` will highly depend on the availability of the service. From a testability perspective, we would rather have our test code inject its own service as a dependency. This *service* would then return an expected result enabling us to verify that the code above as such is working fine. How to do this exactly though? Using an abstraction of the `VATRegistrationValidation` method and the related fields `Country/Region Code` and `VAT Registration No.` – as we cannot alter the standard code. We will discuss this in detail in *Chapter 13, Testing Incoming and Outgoing Calls*, which is dedicated to testing externally accessible components and how to mock them.

> **Note**
> Even though there is more to be said about the testability of the rest of the code of
> `VATRegistrationValidation`, we will stop here as our point regarding **DI** has
> been made.

Not all that glitters is gold

Any technique we apply has its pros and cons. The same applies to the three coding techniques
we discussed here. As noted in the example of dependency removal, the upsides were better
control and faster execution when testing, while the downsides would be the probable retrieval
of the general ledger setup before each call to the refactored method. Where the upsides of **DI**
are similar, a major downside might be that the dependency is abstracted to a higher level and
even moved into configuration. This could bring extra complexity in testing. So, when applying
any of these coding techniques, always ask yourself whether one solution fits better than the
other before deciding what to choose.

Some notes on refactoring your code so it is testable

Unit testing has for a very long time not been the focus in Business Central development. As a
consequence, extracting the *implementation of business logic into testable methods* was hardly
considered. Now, moving forward with your existing code and getting it *efficiently and effectively*
tested, the coding techniques discussed in this section give some important hints on how you
could refactor your code. But always bear in mind to *never change a winning team*, meaning that:
do not refactor code if there is no external reason for it that will guarantee your changes to be
tested; refactoring for the sake of refactoring is never a good reason. Note that external reasons
might range from a bug report to a feature extension and more. Those are planned including the
testing needed to be done.

When having an external reason, take the coding techniques into account to make your code –
better – testable. Note that this will also contribute to a better state of your application code as it
will be more loosely coupled. You might want to study **SOLID** more closely, right?

Make it a joined, team effort to add testability to your code review. And of course, not only when
refactoring, but also when starting from scratch. In the next test example, we'll work out one
of the coding techniques discussed above: *Extract implementation of business logic into testable
methods*, and more specifically, doing this using event subscribers.

Test example 11 – how to unit test event subscribers' logic

In the **LookupValue** extension, we extended the standard application logic by means of the following six event subscribers:

1. `OnCreateCustomerFromTemplateOnBeforeCustomerInsert` Event

   [to be found in `ContactEvents.Codeunit.al`]

2. `OnApplyTemplateOnBeforeCustomerModifyEvent`

   [to be found in `CustomerTemplEvents.Codeunit.Codeunit.al`]

3. `OnAfterCopySellToCustomerAddressFieldsFromCustomer` Event

   [to be found in `SalesHeaderEvents.Codeunit.al`]

4. `OnBeforeCreateShptLineFromSalesLineEvent`

   [to be found in `WhseCreateSourceDocumentEvents.Codeunit.al`]

5. `OnBeforePostSalesDocEvent`

   [to be found in `SalesPostEvents.Codeunit.al`]

6. `OnBeforePostSourceDocumentEvent`

   [to be found in `WhsePostShipmentEvents.Codeunit.al`]

Customer wish

These event subscribers were based on the following customer wishes:

- Event subscribers 1 and 2:

  When creating a customer from a customer template, the `Lookup Value Code` field should be inherited from the `Customer Template` to the `Customer`.

- Event subscriber 3:

  When selecting a customer in the `Sell-to Customer` field on a sales document, the `Lookup Value Code` field should be inherited from the `Customer` to the `Sales Header`.

- Event subscriber 4:

  When creating a warehouse shipment from a sales order, the `Lookup Value Code` field should be inherited from the `Sales Header` to the `Warehouse Shipment Line`.

- Event subscriber 5:

  When posting a sales document, it is mandatory that the Lookup Value Code field is populated.

- Event subscriber 6:

  When posting a warehouse shipment, the Lookup Value Code field should be inherited from the Warehouse Shipment Line to the Posted Whse. Shipment Line.

Application code

With this we created six units to be tested, each of them doing one specific thing:

- The code in the first four subscribers assures that the Lookup Value Code field on the relevant entity is populated by inheritance, like, for example, OnAfterCopySellToCustomerAddressFieldsFromCustomer Event:

```
[EventSubscriber(
  ObjectType::Table, Database::"Sales Header",
  'OnAfterCopySellToCustomerAddressFields
    FromCustomer', '', false, false)]
local procedure
  OnAfterCopySellToCustomerAddressFieldsFrom
    CustomerEvent(
    var SalesHeader: Record "Sales Header";
    SellToCustomer: Record Customer)
begin
  SalesHeader."Lookup Value Code" :=
  SellToCustomer."Lookup Value Code";
end;
```

- The latter two subscribers check, before posting, whether the relevant Lookup Value Code field is populated, like, for example, OnBeforePostSalesDocEvent:

```
[EventSubscriber(
  ObjectType::Codeunit, Codeunit::"Sales-Post",
    'OnBeforePostSalesDoc', '', false, false)]
local procedure OnBeforePostSalesDocEvent(
  SalesHeader: Record "Sales Header")
begin
  SalesHeader.TestField("Lookup Value Code");
end;
```

As discussed in the section *Using event subscribers as a means to isolate testable units*, best practice is, in compliance with **SRP**, to implement each business rule in its own public method. Testing the business logic triggered by an event subscriber then comes down to (unit) testing the relevant public method. So, before working on these units tests, the business logic that so far resided in the event subscribers has to be extracted into testable method. To see the result, have a look on GitHub at our six codeunits that are holding the event subscribers:

- codeunit ContactEvents (50003)
- codeunit CustomerTemplEvents (50005)
- codeunit SalesHeaderEvents (50000)
- codeunit SalesPostEvents (50004)
- codeunit WhseCreateSourceDocumentEvents (50002)
- codeunit WhsePostShipmentEvents (50001)

Test code

So far, we only tested those six new business rules by means of functional tests for (sub) features **Inheritance**, **Posting**, and **Warehouse Shipment** – see, among others, *Chapter 6, From Customer Wish to Test Automation – the Basics*, which does make a lot of sense. Given the above discussed extracting of the business logic, we could, however, also have tested these units on their own. Have a look at the scenarios:

```
[FEATURE] LookupValue UT Posting
[SCENARIO #0100] Check failure Posting
[GIVEN] Sales header without lookup value
[WHEN] Trigger CheckLookupvalueExistsOnSalesHeader
       Sales Posting
[THEN] Missing lookup value on sales header error thrown
[SCENARIO #0101] Check success
               CheckLookupvalueExistsOnSalesHeader
               Sales Posting
[GIVEN] Sales header with lookup value
[WHEN] Trigger CheckLookupvalueExistsOnSalesHeader
       Sales Posting
[THEN] No error thrown
[SCENARIO #0102] Check failure
               CheckLookupvalueExistsOnSalesHeader
               Whse. Posting
```

[GIVEN] Sales header with number and without lookup value
[WHEN] Trigger CheckLookupvalueExistsOnSalesHeader
       Whse. Posting
[THEN] Missing lookup value on sales header error thrown
[SCENARIO #0103] Check success
                 CheckLookupvalueExistsOnSalesHeader
                 Whse. Posting
[GIVEN] Sales header with number and lookup value
[WHEN] Trigger CheckLookupvalueExistsOnSalesHeader
       Whse. Posting
[THEN] No error thrown

[FEATURE] LookupValue UT Inheritance
[SCENARIO #0104] Check InheritLookupValueFromCustomer
[GIVEN] Customer with Lookup value
[GIVEN] Sales header without lookup value
[WHEN] Trigger InheritLookupValueFromCustomer
[THEN] Lookup value on sales document is populated
       with lookup value of customer
[SCENARIO #0105] Check
                 ApplyLookupValueFromCustomerTemplate
                 from Contact
[GIVEN] Customer template with lookup value
[GIVEN] Customer
[WHEN] Trigger ApplyLookupValueFromCustomerTemplate
[THEN] Lookup value on customer is populated with
       lookup value of customer template
[SCENARIO #0106] Check
                 ApplyLookupValueFromCustomerTemplate
[GIVEN] Customer template with lookup value
[GIVEN] Customer
[WHEN] Trigger ApplyLookupValueFromCustomerTemplate
[THEN] Lookup value on customer is populated with
       lookup value of customer template

[FEATURE] LookupValue UT Warehouse Shipment

[SCENARIO #0107] Check
                      InheritLookupValueFromSalesHeader

[GIVEN] Sales header with lookup value

[GIVEN] Warehouse shipment line

[WHEN] Trigger InheritLookupValueFromSalesHeader

[THEN] Lookup value on warehouse shipment line is
          populated with lookup value of sales header

These scenarios have been implemented in the following `.al` test codeunits:

- codeunit 81025 – LookupValue UT Posting

  [to be found in `LookupValueUTPosting.Codeunit.al`]

- codeunit 81026 – LookupValue UT Inheritance

  [to be found in `LookupValueUTInheritance.Codeunit.al`]

- codeunit 81023 – LookupValue UT Whse Shipment

  [to be found in `LookupValueUTWarehouseShipment.Codeunit.al`]

They have been added to `https://github.com/PacktPublishing/Automated-Testing-in-Microsoft-Dynamics-365-Business-Central-Second-Edition/tree/main/Chapter 12 (LookupValue Extension)` on the GitHub repo. Go there to study the details.

Notes

(1) You can find these ATDD scenarios in a separate ATDD sheet in the GitHub repo: `https://github.com/PacktPublishing/Automated-Testing-in-Microsoft-Dynamics-365-Business-Central-Second-Edition/blob/main/Excel sheets/ATDD Scenarios/LookupValue - Chapter 12.xlsx`. This is done to not interfere with the previous test examples.

(2) You might notice that the numbering of the scenarios starts at `#0100`. The reason behind this is that it clearly discriminates these scenarios from the ones created and worked on so far. The numbering of the scenarios is to have a simple and unique reference to each scenario and not so much about having a consecutive series.

Test execution

Now let's add them to the AL Test Tool, and get them running: green!

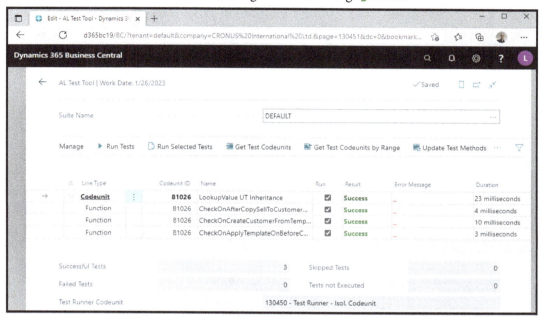

Figure 12.3 – Test example 11 having been run for LookupValue UT Inheritance

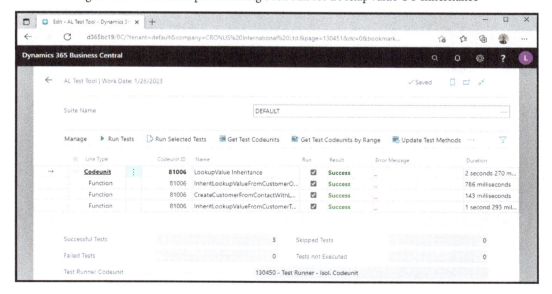

Figure 12.4 – LookupValue Inheritance full scenario (test example 6) run

Notice that how fast these tests are (*Figure 12.3*). It's only a matter of a few milliseconds per test, while the full scenarios, that did trigger the same subscribers, would take much more time, going to about 1 or 2 seconds in the first run (*Figure 12.4*):

Summary

In this chapter, you learned what testable code is. We discussed how you can make your application code testable, by extracting the implementation of business logic into testable methods, reducing the number of code execution paths in each method, and replacing dependencies.

In the next and final chapter, *Chapter 13, Testing Incoming and Outgoing Calls*, we will discuss how to test incoming and outgoing calls to and from external components. We will apply mocks to make tests, that verify business logic interacting with external components, robust. With the latter, we will illustrate **dependency injection**, the technique we talked about in this chapter.

Further reading

As this chapter was an endeavor in bringing together a lot of different and related knowledge, experiences, and skills, I thankfully made use of various resources out there on the internet. Some can be found inline in this chapter. Here are two more:

- On testable code:

  `https://www.toptal.com/qa/how-to-write-testable-code-and-why-it-matters`

- On interface-based **dependency injection**:

  `https://www.c-sharpcorner.com/article/how-to-write-testable-code-in-net/`

13
Testing Incoming and Outgoing Calls

In the connected world that we are in, chances are big that your extension enables external systems to call into Business Central by providing one or more APIs and/or makes its own outgoing calls to external services. Both *incoming* and *outgoing* calls are two mirrored sides of the same medal, where a so-called *client* process is calling into a *service*. With an incoming call, it's our service being called upon by an external client, whereas with an outgoing call, it's an external service being called upon by our client process. Testing the one – handling an incoming call – is about verifying the validity of the service, while testing the other – handling an outgoing call – is about verifying the validity of the client process. To test the incoming call, we need to mimic the calling client; to test the outgoing call, we need to mimic the service being called. In both cases, our challenge is how to do the mimicking of either the client or service.

In this chapter, we will discuss a number of best practices regarding:

- Testing incoming calls
- Testing outgoing calls

With each topic, we will provide a comprehensive test example, *Test example 12 – testing incoming calls: Lookup Value API*, and *Test example 13 – VAT Registration No. validation*.

Technical requirements

Like in the previous chapter, we will refer to the **LookupValue** extension. Its code is to be found on GitHub: `https://github.com/PacktPublishing/Automated-Testing-in-Microsoft-Dynamics-365-Business-Central-Second-Edition`. In the first test example of this chapter, *Test example 12 – testing incoming calls: Lookup Value API*, we will continue on the application code from *Chapter 12, Writing Testable Code*: `https://github.com/PacktPublishing/Automated-Testing-in-Microsoft-Dynamics-365-Business-Central-Second-Edition/tree/main/Chapter 12 (LookupValue Extension)`.

The final code of this chapter can be found in `https://github.com/PacktPublishing/Automated-Testing-in-Microsoft-Dynamics-365-Business-Central-Second-Edition/tree/main/Chapter 13 (LookupValue Extension)` for *Test example 12 – testing incoming calls: Lookup Value API* and `https://github.com/PacktPublishing/Automated-Testing-in-Microsoft-Dynamics-365-Business-Central-Second-Edition/tree/main/Chapter 13 (VAT Registration No. Validation)` for *Test example 13 – VAT Registration No. validation*.

Details on how to use this repository and how to set up VS Code and the Business Central environment are discussed in *Appendix, Getting Up and Running with Business Central, VS Code, and the GitHub Project*.

Testing incoming calls

Connecting from an external system to Business Central has been enabled in Business Central by so-called *Web Services* and *APIs*. Where web service, as such, is a very generic term, entailing both *Web Services* and *APIs*, Business Central has used it to name their very first implementation of web services after it. *Web Services* were introduced in Dynamics NAV 2009 and are still available in Business Central. They can either be SOAP- or OData-based. Given a number of limitations of *Web Services*, however, Microsoft has put in place a new version of web services, which are called **APIs** and are **REST**-compliant, a.k.a. **RESTful**. In his Areopa webinar *Working with APIs*, Arend-Jan Kauffmann gives a nice introduction to APIs starting with a comparison between Business Central's *Web Services* and *APIs*. If APIs are new to you, you can learn more about it with the recording of the aforementioned webinar available at: `https://www.youtube.com/watch?v=-BLNMTf7r5k`.

> **Note**
> Arend-Jan Kauffmann, a.k.a. AJ, is, in my humble opinion, the go-to expert on web services in the Business Central world. You will find a lot more info on working with Web Services (`https://www.kauffmann.nl/category/webservices/`) and APIs (`https://www.kauffmann.nl/category/api/`) on his blog.

As APIs are the way to go for your future implementation of web services in Business Central, we will talk about how to test your own Business Central API in this section. Or as stated in the introduction: how to verify the validity of your service and mimic a client calling it. We will tackle a number of technical challenges and show you – given well-defined ATDD scenarios as always – how you can fairly easily set up your test code by making use of the standard test library `Library - Graph`.

Technical challenges

Testing an API from a high level is like any other test flow:

1. Set up the fixture – GIVEN.

2. Exercise the action – WHEN.

3. Verify the result – THEN.

No apparent complexity as such. But executing an API call always involves two different processes: the calling or invoking side, a.k.a. the *client*, and the receiving or invoked side, a.k.a. the *server*. In this context, there are three technical challenges:

*   How to *control the test flow* while the execution takes place in two different processes – the client and the server?

*   How to handle the *authentication of the client* to the server before the latter can be invoked?

*   How to *set up the client-server interaction*?

Controlling the test flow

Where in real life, client and server will run on two different systems, handling an API call in a similar test setup induces a certain complexity. Some parts do happen on the client-side, while the others happen server-side:

*   Setting up and exercising the REST request and checking its response happens *client-side*.

*   Getting the data set up before the REST request and verifying it – after the action – happens *server-side*.

In a Business Central automated test, we will – quasi – switch between client-side and server-side in the following manner as we make sure that both test code and API reside in the same environment:

1. GIVEN: set up the *server-side* data and prepare the *client-side* REST request.

2. WHEN: exercise the action, that is, set up and send the *client-side* REST request, and receive the response.

3. **THEN**: verify *server-side* the resulting changes in the database and verify *client-side* the resulting REST request-response.

Have a look at *Figure 13.1* for a schematic representation of this test flow:

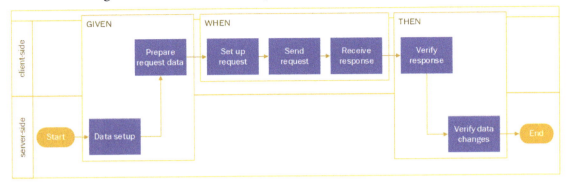

Figure 13.1 – Incoming REST call test flow

Note that the server-side response is not an explicit part of the test flow and, as such, not shown in the flow chart. It defines, next to the data setup for and verification of the server, how to mimic the calling client. In addition to this, we have to realize that this test flow is executed in one session, while the invoked server-side process will run in another session and both in the same environment. To enable the invoked server-side process to pick up the data set up in the **GIVEN** part, this data should be committed to the database. This means that our test cannot run in isolation to assure that a commit to the database will be a real commit. Hence, API tests need to run with test isolation disabled. This can be easily achieved fortunately by using the standard test runner `Test Runner - Isol. Disabled` (130451).

> **Note**
>
> If you want to be able to rerun your tests directly after each other without refreshing the database, make sure your tests are set to do that. In other words, your tests should probably not reuse the same data each run but should make use of fresh data. You accomplish this by ensuring that records have unique primary key values assigned.

Authenticating the client

The next challenge is the authentication of the client as the service being called upon will require the client's credentials. Including the authentication in tests would be quite cumbersome, and therefore the Microsoft Business Central team has chosen the practical approach, focusing on testing the API business logic and not validating the authentication. As such API tests are run in a database with no user and user-related data. In order to achieve this, the following needs to be done:

1. Set up your Business Central environment to use Windows authentication.
2. Remove all user and user-related data from the database.
3. Deploy your extension(s) with your application and test code.
4. Run the tests with standard test runner `Test Runner - Isol. Disabled` (130451).

Note that with this approach, we ignore testing the authorization to the server.

Now, having no user (related) data in the database will not trigger the user authentication, but allow the client session to log in. To achieve this as per *step 2* above, there are a number of different ways:

- After installing Business Central from either the product DVD or as a container, run the following SQL script on the relevant database:

```
USE [your-bc-database-name]
GO


DELETE FROM [dbo].[User]
DELETE FROM [dbo].[User Property]
DELETE FROM [dbo].[Page Data Personalization]
DELETE FROM [dbo].[User Personalization]
DELETE FROM [dbo].[Access Control]
GO
```

Note that the highlighted part needs to get replaced by your specifics. You can find a somewhat extended script in the GitHub repo: `https://github.com/PacktPublishing/Automated-Testing-in-Microsoft-Dynamics-365-Business-Central-Second-Edition/blob/main/Scripts/RemoveAllUserRelatedData.sql`.

- Create a docker container with Windows authentication, valid Windows credentials, but no user in the database. To get the latter done, add the following parameter to the **BcContainerHelper** PowerShell cmdlet `New-BcContainer`:

  ```
  -myscripts @( @{ "SetupNavUsers.ps1" = "Write-Host 'Skipping
  user creation'" } )
  ```

- Restoring a standard database backup from either the product DVD or the artifacts as this does not contain any users and user-related data yet.

> **Notes**
>
> (1) In all cases, the authentication of Business Central has to be Windows. When deploying from VS Code, valid Windows credentials should apply for the user that is doing the deployment. In case of a docker container, you have to ensure that you're connected to the Windows domain. The latter is not needed when deploying against a local environment installed from the product DVD.
>
> (2) When installing Business Central from the product DVD, its service tier will not have the OData and API services enabled. Turn these on and restart the service tier before you run your tests.

Setting up the client-server interaction

The third challenge is to set up the client-server interaction, which is sending the REST request from the client to the server. For this, we need to do the following:

1. Set up endpoint or target URL, including a reference to the company addressed.
2. Set the request method, either GET, DELETE, POST, or PATCH.
3. Set the request content and return type.
4. Execute the request.

What would that look like given a new API called `lookupValue`, which we will discuss in more detail in *Test example 12 – testing incoming calls: Lookup Value API*?

Calling a supposed API `lookupValue` using the VS Code **REST client** to retrieve the first lookup value record could look like:

```
#### Get & store all companies for API publisher/group ###
# @name companies
get {{baseurl}}/api/fluxxus/automation/v1.0/companies?
  tenant=default
```

```
Authorization: Basic {{username}}:{{password}}


#### Get all lookup values for first company #########
# @name lookupvalues @companyid = {{companies.response.body.
value[0].id}}
get {{baseurl}}/api/fluxxus/automation/v1.0/
   companies({{companyid}})/lookupValues?tenant=default
Authorization: Basic {{username}}:{{password}}


#### Get first lookup values for first company #######
@lookupvalueid =
   {{lookupvalues.response.body.value[0].id}}
get {{baseurl}}/api/fluxxus/automation/v1.0/
   companies({{companyid}})/
   lookupValues({{lookupvalueid}})?tenant=default
Authorization: Basic {{username}}:{{password}}
```

You might give it a try given you have installed the **LookupValue** extension as per the code on GitHub for this chapter and the REST client for VS Code. The full http request listing can be found in `https://github.com/PacktPublishing/Automated-Testing-in-Microsoft-Dynamics-365-Business-Central-Second-Edition/blob/main/Chapter 13 (LookupValue Extension)/http.requests/test.http`.

Although using a REST client is a very handy way to interact with a restful API, this, however, is not the way to address the API from AL code. Our code should set up the right http request and, of course, after the request, receive, read, and check its response. All together quite some coding and I am lucky – and I reckon you are too – Microsoft has done this job for us already. In `Library - Graph`, they have brought together a whole bunch of reusable functions for handling restful API calls, which we will use in our next test example: *Test example 12 – testing incoming calls: Lookup Value API.*

Notes

(1) Find the REST Client extension for VS Code here: `https://marketplace.visualstudio.com/items?itemName=humao.rest-client`.

(2) Graph in `Library - Graph` relates to Microsoft Graph, a unified REST API by Microsoft that gives you access to a giant collection of data in the cloud-like Dynamics 365. Read more on MS Graph here: `https://docs.microsoft.com/en-us/graph/overview` or `https://www.poweronplatforms.com/microsoft-graph-what-is-it/`.

Examples of standard API tests

Before we go and elaborate on our text example, let's mention a number of standard test examples, leaving it to you to go and study the details yourself.

Both the `_Exclude_APIV1_Tests` and `_Exclude_APIV2_Tests` apps contain a lot of illustrative examples of API tests. As a matter of fact, I heavily made use of them. As their structure is in essence – almost – every time the same, you could say that any of the test codeunits you do find in these two test apps will be of great help to create your own API tests. Let's take out two good example codeunits:

- `Codeunit 139722 - APIV1 - Employees E2E` in `_Exclude_APIV1_Tests` app

  It contains four test functions that check the `GET`, `DELETE`, `PATCH`, and `POST` calls to the standard API page `APIV1 - Employees`: `TestGetEmployee`, `TestDeleteEmployee`, `TestModifyEmployee`, and `TestCreateEmployee`.

- `Codeunit 139700 - APIV1 - Items E2E` in `_Exclude_APIV1_Tests` app

  This test codeunit contains a lot more test functions that check the `GET`, `DELETE`, `PATCH`, and `POST` calls to the standard API page: `APIV1 - Items`. Let's name just the four basic ones: `TestGetSimpleItem`, `TestDeleteItem`, `TestModifyItem`, and `TestCreateSimpleItem`.

> **Note**
>
> As the name of the `_Exclude_APIV1_Tests` and `_Exclude_APIV2_Tests` apps indicate, the first contains the version 1.0 implementation of standard APIs, while the latter the version 2.0 implementation. Their sources can be found in the `Applications` folder on the DVD, in the artifacts feed, or in the GitHub repo, `https://github.com/microsoft/ALAppExtensions`, in Apps/W1.

Testing outgoing calls

In everyday practice, both incoming and outgoing calls happen between two separate, independent systems. This makes testing of incoming and outgoing calls a challenge with respect to controlling the flow of the tests. As shown in the previous section, *Testing incoming calls*, with incoming calls we can quasi switch between client-side and server-side while actually running our tests on one system. With outgoing calls, however, this is not possible as our application code – the *client* – is most likely calling upon an external system – the *server* – that we do not and cannot control and/or we do not want to interact with for whatever reason in our test context. In *Chapter 12, Writing Testable Code*, handling the interaction with an external system was mentioned as an example of why we should be writing testable code. Testable code should allow us to mimic the interaction and test the behavior of our application code given a controllable response of the external system. This is what often is called **mocking** but more generically is known as **test double**. Now, before we focus on testing outgoing calls, let's first elaborate a bit more on the test doubles terminology.

Test Doubles

Test doubles are components that are used as *stand-ins* to real components – or service/servers as we have called them in the context of this chapter and *Chapter 12*. You might know them as **dummy**, **stub**, **fake**, **spy**, or **mock**. The term **test double** was introduced by Gerard Meszaros, the writer of *XUnit Test Patterns*, as a parallel to stunt doubles in movies, replacing the real actor to mimic this actor's behavior in a specific, often dangerous, scene. As a **mock** is actually a very specific **test double** it makes a lot of sense to use the generic term **test double** and distinguish between the specific test doubles and how to build and apply them. In the context of this book, however, now introducing the concept of using stand-in objects to simulate components, we will use the generic term **test double** without discussing the specific types. The distinction becomes important once your test environment becomes more sophisticated, something to address in the future.

> **Notes**
>
> (1) Refer to this web page `http://xunitpatterns.com/Test Double Patterns.html` to read what Gerard Meszaros has to say about test doubles in his very structured way of describing things. Personally, I prefer how Martin Fowler talks about test doubles and their different forms in his post at `https://martinfowler.com/articles/mocksArentStubs.html`, in the *Difference between Mocks and Stubs* section (mind you that the whole article is more than worth reading). Especially his clear definitions of the different test doubles, **dummy**, **stub**, **fake**, **spy**, or **mock**, as there seems to be quite opposing descriptions of some of these terms as, for example, this discussion on stack overflow shows: `https://stackoverflow.com/questions/346372/whats-the-difference-between-faking-mocking-and-stubbing`.

> (2) Related to *Chapter 2, Test Automation and Test-Driven Development*, but out of scope of this chapter, it is worthwhile pointing to the last part of Martin Fowler's article referred to above on *classical and mockist testing*.

Replacing a server with a test double – dependency injection

Here, *Chapter 12* and *Chapter 13* come together: how can we replace a real component, an external server, with a test double? By **dependency injection** (**DI**)!

As described in *Chapter 12*, the goal of **DI** is to …

…move the instantiation of a dependency – referred to as service – outside of the unit – referred to as client.

Where **DI** is a general software design principle that gives you the control over, as well as flexibility in using, the right service, it can be applied to both production and test environments. This allows you to inject *the right server for the right context at the right time.*

Now, let's pick up the thread we left in *Chapter 12*: the VATRegistrationValidation method, our **DI** example. As discussed there, given that the EU VAT Registration No. Validation Service Setup has been configured and enabled, the VATRegistrationValidation method will – indirectly – make a call on the service defined in the **Service Endpoint** on the **VAT Registration No. Validation Service Setup** page. Testing that a VAT registration number gets validated by the service needs that service to be up and running and reachable by our call. To test our business logic that leads to calling that validation service and receiving its response, we do not really need that specific service to respond. We could also do, and even do better, with a call to and response from a test double of the service, a stub or fake component we fully control ourselves.

Figure 13.2 shows the high-level flow of the validation of a VAT registration number as, for example, when creating a new EU customer in Business Central. Once both a country/region code and a VAT registration number have been provided, the preliminary checks are done and the so-called VAT Registration Log has been initialized, the validation service will be called, its response captured, and processed into the registration log:

Figure 13.2 – VAT Registration No. validation flow with a test double server

As also depicted in *Figure 13.2*, the service could be replaced by a test double using **DI**. How to go about this? We will work this out in more detail in *Test example 13 – VAT Registration No. validation*.

Some more notes on mock and mocking

As the term **mocking** is widely used, where it actually is targeting the wider collection of **test doubles**, this might raise some confusion. Now, be prepared to confront this confusion even; be prepared to find that in many Microsoft test libraries, the term mock is used in the same way and also in my code in *Test example 13 – VAT Registration No. validation* in order to stay close to the terminology in standard code.

In the next section, *Examples of standard tests with test doubles replacing external components*, a number of examples of the standard tests using test doubles – called mocks ;-) – are discussed shortly.

You might also notice that in a lot of standard test codeunits, so-called mocks are also defined as test doubles for data. These are shortcuts to create data-like entries that normally need a posting to be exercised or document lines with no context. Have a look, for example, at the `Library - Sales` codeunit in the `Tests-Libraries` app, and methods such as `MockCustLedgerEntry` and `MockApplnDetailedCustLedgerEntry`.

Examples of standard tests with test doubles replacing external components

In this section, you can find a number of examples of standard test codeunits with test doubles to replace other components:

- `Codeunit 139760 - SMTP Connector Test in Email - SMTP Connector Tests` app

  Example test method: `TestSendMailPayloadMessage`

  This test method makes use of the fact that the `SMTP Client` has been implemented as an interface and for this specific test, an SMTP client test double is injected to mimic the handling of the email setup and sending of the email.

- `Codeunit 139751 - Microsoft 365 Connector Tests in Email - Microsoft 365 Connector Tests` app

  Example test method: `TestSendMailPayloadMessage`

  In a similar way as the previous example, this test method also makes use of the fact that `Outlook API Client` and `OAuth Client` have been implemented both as interfaces. The injection happens through a subscriber that resides in a codeunit that needs to "manually" be bound through the `BindSubscription` method.

Test example 12 – testing incoming calls: Lookup Value API

With this example, you learn how to test getting (using `GET`), deleting (using `DELETE`), modifying (using `PATCH`), and creating (using `POST`) data through an API.

> **Note**
>
> In this example, we focus on the structure and code of the automated tests that test an API, not on the flow of getting from a customer wish into test and application coding and also not on doing it the *TDD way*.

Application code

The object for our tests is a new page called `Lookup Values APIV1` (50090) with `PageType` set to `API` and `SourceTable` pointing to our `Lookup Value` table. This API contains three columns:

- `id` based on `SystemId` field
- `number` based on the `Code` field
- `displayName` based on the `Description` field

The full object can be seen in the GitHub repo: `https://github.com/PacktPublishing/Automated-Testing-in-Microsoft-Dynamics-365-Business-Central-Second-Edition/blob/main/Chapter 13 (LookupValue Extension)/src/page/LookupValuesAPIV1.Page.al`.

> **Note**
>
> If you have missed out on this, the `SystemId` field is one of the so-called system fields introduced in Business Central 2019 wave 2 – BC15 – because of integration with external system. Read more on system fields here: `https://docs.microsoft.com/en-us/dynamics365/business-central/dev-itpro/developer/devenv-table-system-fields`.

Test scenarios

To verify getting (GET), deleting (DELETE), modifying (PATCH), and creating (POST) data through an API, we need the following four scenarios:

```
[FEATURE] LookupValue UT API
[SCENARIO #0200] Get lookup value
[SCENARIO #0201] Delete lookup value
[SCENARIO #0202] Modify lookup value
[SCENARIO #0203] Create lookup value
```

In addition to this, two more tests are needed as our API does not allow for an empty lookup value description – I suggest you have a look at the API page Lookup Values APIV1 (50090) to see why:

```
[SCENARIO #0204] Modify lookup value with empty description
[SCENARIO #0205] Create lookup value with empty description
```

Let's pick out two of these six scenarios to elaborate on: #0200 and #0202. I'll leave it to you to study the other four given the code in the GitHub repo.

Before we go and code these scenarios, their GIVEN-WHEN-THEN details need to be provided:

```
[SCENARIO #0200] Get lookup value
[GIVEN] Committed lookup value
[WHEN] Send GET request for lookup value
[THEN] Lookup value in response
[SCENARIO #0202] Modify lookup value
[GIVEN] Committed lookup value
[GIVEN] Updated lookup value JSON object
[WHEN] Send PATCH request for lookup value
[THEN] Updated lookup value in response
[THEN] Updated lookup value in database
```

To relate to the generic test flow as discussed in *Controlling the test flow*, *Figure 13.3* and *Figure 13.4* display the test flow schemas for scenarios #0200 and #0202, respectively:

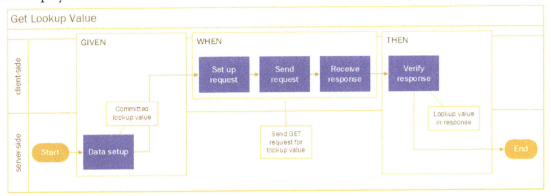

Figure 13.3 – Scenario #0200: "Get Lookup Value" test flow

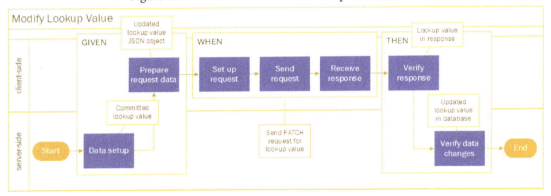

Figure 13.4 – Scenario #0202: "Modify Lookup Value" test flow

You might notice that the *Get Lookup Value* test flow does not need all the parts for GIVEN and THEN.

Test Code – [SCENARIO #0200] Get lookup value

Performing in one go our 4-steps recipe on scenario #0200 leads to this test function:

```
[Test]
procedure GetLookupValue()
var
    LookupValue: Record LookupValue;
```

```
    Response: Text;
begin
    // [SCENARIO #0200] Get lookup value
    // [GIVEN] Committed lookup value
    CreateCommittedLookupValue(LookupValue);
    // [WHEN] Send GET request for lookup value
    Response := SendGetRequestForLookupValue(LookupValue);
    // [THEN] Lookup value in response
    VerifyLookupValueInResponse(Response, LookupValue);
end;
```

Introducing the following helper functions:

- `CreateCommittedLookupValue`

- `SendGetRequestForLookupValue`

- `VerifyLookupValueInResponse`

Let's discuss their details unveiling some API testing specifics.

CreateCommittedLookupValue

With `CreateCommittedLookupValue`, the *server-side* data setup is taken care of by creating a new lookup value. Here, our standardized `CreateLookupValue` is called, which we supplement with a `Commit` in order to unlock tables and allow the API web service – running in a separate session – to pick up changes:

```
local procedure CreateCommittedLookupValue(
    var LookupValue: Record LookupValue)
var
    LibraryLookupValue:
        Codeunit "Library - Lookup Value";
begin
    LibraryLookupValue.CreateLookupValue(LookupValue);
    Commit();
end;
```

SendGetRequestForLookupValue

Moving *client-side*, `SendGetRequestForLookupValue` will set up and send the GET request and also receive the response, making use of the `GetFromWebService` method in the standard test library `Library - Graph`. Using this standard method facilitates a fairly easy development effort as we only need to set up and hand over the target URL for the lookup value we want to get:

```
local procedure SendGetRequestForLookupValue(
  LookupValue: Record LookupValue) Response: Text
begin
  LibraryGraphMgt.GetFromWebService(
    Response, CreateTargetURL(LookupValue.SystemId));
end;
```

The target URL to be handed over is created by a local `CreateTargetURL` method, wrapping another method in the standard test library, `Library - Graph`, with the same name. We only need to provide to this method the `SystemId` of the lookup value record we want to get, the ID of the API page, and the name of the entity we're addressing. Have a look:

```
local procedure CreateTargetURL(ID: Text): Text
begin
  exit(
    LibraryGraphMgt.CreateTargetURL(
      ID, Page::"Lookup Values APIV1",'lookupValues'));
end;
```

VerifyLookupValueInResponse

Now that the server has been requested to get the lookup value and the response has been received, the only thing remaining is a *client-side* verification of the response by means of `VerifyLookupValueInResponse`:

```
local procedure VerifyLookupValueInResponse(
  JSON: Text; LookupValue: Record LookupValue)
begin
  Assert.AreNotEqual('', JSON,
    'JSON should not be empty.');
  LibraryGraphMgt.VerifyIDInJson(JSON);
  LibraryGraphMgt.VerifyPropertyInJSON(
    JSON, 'number', LookupValue.Code);
  LibraryGraphMgt.VerifyPropertyInJSON(
```

```
        JSON, 'displayName', LookupValue.Description);
end;
```

Making use of another set of standard methods, `VerifyLookupValueInResponse` checks whether the JSON response:

- Is not empty,
- Contains an ID part, and
- Contains the lookup value record identified by `number` and `displayName`.

Now that we did get our first and simplest API test coded, we're ready to step into the next one that entails somewhat more, as scenario `#0202` needs some more preps – GIVEN – and verifications – THEN.

Test Code – [SCENARIO #0202] Modify lookup value

Again performing in one go, our 4-steps recipe on scenario `#0202` delivers us test function `ModifyLookupValue`:

```
[Test]
procedure ModifyLookupValue()
var
  LookupValue: Record LookupValue;
  TempLookupValue: Record LookupValue temporary;
  RequestBody: Text;
  Response: Text;
begin
  // [SCENARIO #0202] Modify lookup value
  // [GIVEN] Committed lookup value
  CreateCommittedLookupValue(LookupValue);
  // [GIVEN] Updated lookup value JSON object
  RequestBody := CreateUpdatedLookupValueJSONObject(
    LookupValue.Code, TempLookupValue);
  // [WHEN] Send PATCH request for lookup value
  Response := SendPatchRequestForLookupValue(
    LookupValue, RequestBody);
  // [THEN] Updated lookup value in response
  VerifyLookupValueInResponse(Response,TempLookupValue);
  // [THEN] Updated lookup value in database
  VerifyLookupValueInDatabase(LookupValue.Code, Response);
end;
```

With the previous test function for scenario #0200 in mind, together with the schemas in *Figure 13.3* and *Figure 13.4*, you can recognize two common parts, the first GIVEN and first THEN:

- `CreateCommittedLookupValue`
- `VerifyLookupValueInResponse`

Secondly the WHEN part has a resemblance too, but of course it's not exactly the same as it is doing a PATCH request:

- `SendPatchRequestForLookupValue`

The major, explicit, differences are the second GIVEN and second THEN parts:

- `CreateUpdatedLookupValueJSONObject`
- `VerifyLookupValueInDatabase`

There is an extra prep on the client-side and an extra verification on the server-side.

Let's discuss their details to unveil some more API testing specifics.

CreateCommittedLookupValue

`CreateCommittedLookupValue` is the *server-side* data setup as for scenario #0200, discussed already for *Figure 13.3*.

CreateUpdatedLookupValueJSONObject

Wanting to modify the newly created lookup value – stored in `LookupValue` – we need to move *client-side* to set up an updated version of our lookup value – stored in `TempLookupValue`.

To achieve this, `CreateUpdatedLookupValueJSONObject` calls the local method `GetLookupValueJSONObject` that returns as text the JSON formatted version of `TempLookupValue`. Note that for this it is making use of the `AddPropertytoJSON` method in the standard test library `Library – Graph`:

```
local procedure CreateUpdatedLookupValueJSONObject(
  LookupValueCode: Code[10];
  var TempLookupValue: Record LookupValue temporary): Text
begin
  TempLookupValue.Code := LookupValueCode;
  exit(GetLookupValueJSONObject(
    TempLookupValue.Code, TempLookupValue.Code,
```

```
      TempLookupValue));
end;


local procedure GetLookupValueJSONObject(
  NewCode: Code[10]; NewDescription: Text[50];
  var TempLookupValue: Record LookupValue temporary)
  LookupValueJSON: Text
begin
  TempLookupValue.Code := NewCode;
  TempLookupValue.Description := NewDescription;
  LookupValueJSON :=
    LibraryGraphMgt.AddPropertytoJSON(
      LookupValueJSON, 'number', LookupValue.Code);
  LookupValueJSON :=
    LibraryGraphMgt.AddPropertytoJSON(
      LookupValueJSON, 'displayName',
        LookupValue.Description);
end;
```

This JSON object returned by CreateUpdatedLookupValueJSONObject is handed over as the so-called request body to SendPatchRequestForLookupValue, which we will discuss next.

SendPatchRequestForLookupValue

Still on *client-side*, SendPatchRequestForLookupValue will set up and send the PATCH request and also receive the response making use of the PatchToWebService method in the standard test library Library – Graph. Again using a standard method facilitates a fairly easy development effort. In this case, we only need to set up and hand over the target URL for the lookup value we want to modify and hand over the request body generated by CreateUpdatedLookupValueJSONObject:

```
local procedure SendPatchRequestForLookupValue(
  LookupValue: Record LookupValue;
  RequestBody: Text) Response: Text
begin
  LibraryGraphMgt.PatchToWebService(
    CreateTargetURL(LookupValue.SystemId),
      RequestBody, Response);
end;
```

VerifyLookupValueInResponse

The *client-side* verification is similar to that for scenario #0200, as we discussed already for *Figure 13.3*. The major difference here is that the response is checked to contain the updated lookup value – stored in `TempLookupValue` – and not to check the originally created one – stored in `LookupValue`.

VerifyLookupValueInDatabase

And finally, we move *server-side* to verify whether the lookup value description indeed has been updated in the database. As such, we need to retrieve the record from the database and check whether its content is to be found in the response:

```
local procedure VerifyLookupValueInDatabase(
  LookupValueCode: Code[10]; Response: Text)
var
  LookupValue: Record LookupValue;
begin
  LookupValue.Get(LookupValueCode);
  VerifyLookupValueInResponse(Response, LookupValue);
end;
```

Test preparation

Before we can run our tests, we should remind ourselves that we need a specific setup as discussed in the *Authenticating the client* section:

- *Controlling the test flow* explained that API tests should run outside of test isolation, so, we have to make sure that the standard test runner `Test Runner - Isol. Disabled` (130451) is selected in the **AL Test Tool**.

- *Authenticating the client* explained that our Dynamics 365 Business Central environment should have a Windows authentication set and no users (related data) should be present in the database.

On the premise that this has been handled, we are able to run the tests.

> **Note**
> As `Library - Graph` is heavily making use of DotNet components, these tests can only be run in an on-prem environment.

Test execution

Now we're ready to run!

Running in web client

Let's deploy the **LookupValue** extension to our rightly prepared Business Central environment, open the **AL Test Tool**, add `codeunit 81090` to it, and run it.

Recalling that our API feature entails six scenarios, *Figure 13.5* displays all yielding success: `green`!

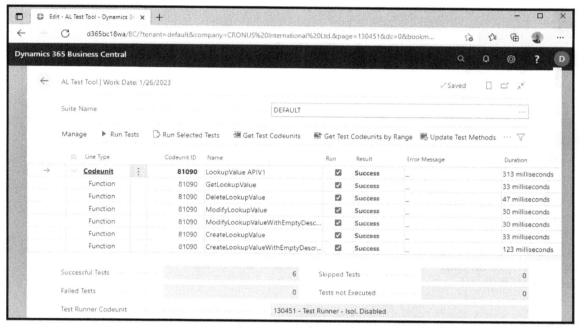

Figure 13.5 – Test example 12 having been run in AL Test Tool in the web client

Running with Run-TestsInBcContainer

We could also trigger a test run from outside using either **AL Test Runner** or with PowerShell using, for example, the **BcContainerHelper** cmdlet `Run-TestsInBcContainer`. You can view its results in VS Code in *Figure 13.6*:

Figure 13.6 – Test example 12 having been run using Run-TestsInBcContainer

As noted in the section *Controlling the test flow*, running the tests with the standard test runner `Test Runner - Isol. Disabled (130451)` will leave the data created by the tests in the database. With the following simple check, you can verify this. If you open up the `Lookup Values` page, you can see that there were indeed records created. Likewise, there is also a new number series with code *GUID*. This one is created and used by the `Library - Utilities`. Have a look at the `No. Series` page to check this too.

Test example 13 – VAT Registration No. validation

In this example, we are going to show how the standard VAT registration number validation feature in Business Central could be made testable, enabling us to mimic the validation service. For your convenience, we have duplicated *Figure 13.2* here, showing you the basic flow of the VAT registration number validation:

Figure 13.7 – VAT Registration No. validation flow with a test double server

As we are not able to modify the code behind this standard feature, and as the code does not provide the right means to hook into it, we will build a standalone feature that imitates the process, as discussed in *Replacing a server with a test double – dependency injection* section. This standalone feature allows us to demonstrate how **DI** could help make an outgoing call testable.

> **Note**
>
> As in the previous example, we will again focus on the structure and code of the automated tests and not on the flow of getting from a *customer wish into test and application coding* and also not on doing it the *TDD way*.

Application code

The basis of this standalone feature is a table called VAT Registration No. holding the most relevant fields to be found on the standard Customer, Vendor, and Contact tables with respect to the VAT registration number validation: Entry No. (mimicking the No. field on Customer, Vendor or Contact tables), Country/Region Code, and VAT Registration No.:

```
field(1; "Entry No."; Code[20])
{
}
field(35; "Country/Region Code"; Code[10])
{
  TableRelation = "Country/Region";

  trigger OnValidate()
  begin
    if "Country/Region Code" <>
      xRec."Country/Region Code"
    then
      VATRegistrationValidation();
  end;
}
field(86; "VAT Registration No."; Text[20])
{
  trigger OnValidate()
  begin
    "VAT Registration No." :=
      UpperCase("VAT Registration No.");
    if "VAT Registration No." <>
```

```
        xRec."VAT Registration No."
    then
        VATRegistrationValidation();
    end;
}
```

As you might recall from *Chapter 12*, the `OnValidate` of both the `Country/Region Code` and `VAT Registration No.` fields holds the standard code necessary to trigger the VAT registration number validation.

To combine the different implementations in this test example in one app, an additional `Enum` field called `Service Handling Type` has been created to enable steering **DI**. So, this allows me to select the original code or the alternative **DI**-enabled code:

```
field(60100; "Service Handling Type";
    Enum "Service Handling Type")
```

Its Enum data type has been defined as follows (The meaning of each option will become clear further down.):

```
enum 60100 "Service Handling Type"
{
    value(0; "Default Codeunit"){}
    value(1; "From Setup"){}
    value(2; "With Subscriber"){}
}
```

As *Figure 13.7* shows, after both the `Country/Region Code` and `VAT Registration No.` have been populated, the formatting of the second field will be checked. This is a generic check that will be done with or without the validation service being called. If the service is enabled, `Country/Region Code` will be checked to establish whether it is in the EU, and if so, the service will be called. The response of the service will be logged in the `VAT Registration Log` table.

> **Note**
> Only the most relevant details are shown in the code above and below. Have a look at the GitHub repo to get the full details. Please refer to the *Technical requirements* section for where to find the code.

The next sequence shows the flow when each step is executed successfully in the UI. If you're going to manually check this yourself, make sure that the **Service Endpoint** is enabled on the **VAT Registration No. Validation Service Setup** page.

1. Enter all fields in the **VAT Registration Nos.** page triggering validation process when leaving **VAT Registration No.** field (*Figure 13.8*):

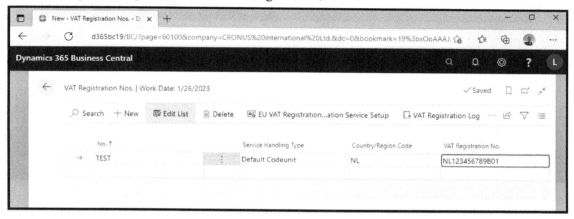

Figure 13.8 – Step 1: Enter all fields, triggering validation process

2. The **Validation Details** page pops up showing the received details for the user to validate (*Figure 13.9*):

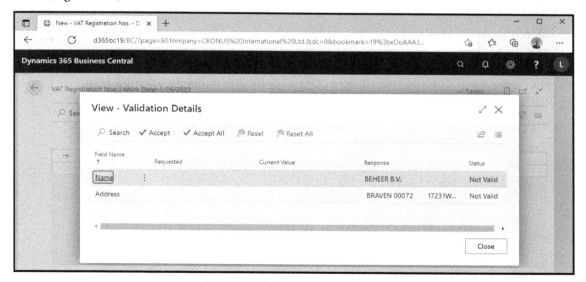

Figure 13.9 – Step 2: Validation details page pops up showing received details

3. When closing the **Validation Details** page, return to the **VAT Registration Nos.** page (*Figure 13.8*) where the user can open the **VAT Registration Log** page to view the resulting **VAT Registration Log** (*Figure 13.10*):

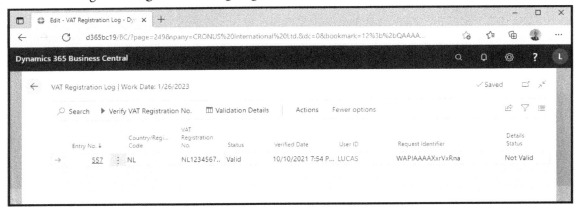

Figure 13.10 – Step 3: View resulting VAT Registration Log

In *step 1* of *Figure 13.8*, it can be seen that the `Service Handling Type` contains the value `Default Codeunit`. This means that the default code – a copy of the same on the `Customer` table – is being executed with a hard-coded reference to the code unit calling the validation service and getting its response. When the service is enabled and available, this should lead to a positive response given a valid VAT registration number, as shown in the next two steps in *Figure 13.9* and *Figure 13.10*.

In the aforementioned section *Replacing a server with a test double – dependency* injection of *Chapter 12*, we saw that the `VATRegistrationValidation` method called from the `OnValidate` trigger of both the `Country/Region Code` and `VAT Registration No.` field calls on the `ValidateVATRegNoWithVIES` method of codeunit `VAT Registration Log Mgt.` (249):

```
if VATRegNoSrvConfig.VATRegNoSrvIsEnabled then begin
  LogNotVerified := false;
  VATRegistrationLogMgt.ValidateVATRegNoWithVIES(
    ResultRecordRef, Rec, "No.",
    VATRegistrationLog."Account Type"::Customer.AsInteger(),
    ApplicableCountryCode);
  ResultRecordRef.SetTable(Rec);
end;
```

The `ValidateVATRegNoWithVIES` method of codeunit `VAT Registration Log Mgt.` (249) in its turn calls on the method `CheckVIESForVATNo` where we find a hard-coded call to the service handling codeunit `VAT Lookup Ext. Data Hndl` (248):

```
Codeunit.Run(
    Codeunit::"VAT Lookup Ext. Data Hndl", VATRegistrationLog);
```

Looking at the context of this call, it is clear indeed that there is no way to inject another dependency:

```
procedure CheckVIESForVATNo(...)
var
    ...
begin
    ...
    if not CountryRegion.IsEUCountry(CountryCode) then
        exit; // VAT Reg. check Srv. is only available for EU
              // countries.

    if VATRegNoSrvConfig.VATRegNoSrvIsEnabled() then begin
        ...
        VATRegNo := VatRegNoFieldRef.Value;

        VATRegistrationLog.InitVATRegLog(
            VATRegistrationLog, CountryCode,
                AccountType, EntryNo, VATRegNo);
        Codeunit.Run(
            Codeunit::"VAT Lookup Ext. Data Hndl",
            VATRegistrationLog);
    end;
end;
```

Therefore, we are going to create example code for how this could be done, hence the two other options in the `Service Handling Type Enum`:

- `From Setup`:

  In this case, we enable **DI** by abstracting the dependency away through a parameter, `VATRegNoSrvCodeunitId`, in all relevant functions defining the ID of the codeunit that will handle the interaction with the service:

  ```
  if VATRegNoSrvCodeunitId <> 0 then
      Codeunit.Run(VATRegNoSrvCodeunitId, VATRegistrationLog)
  ```

```
else
  Codeunit.Run(
    Codeunit::"VAT Lookup Ext. Data Hndl",
      VATRegistrationLog);
```

In our example code, we combine this by having added the ability to provide the codeunit ID on the **EU VAT Registration No. Validation Service Setup** page.

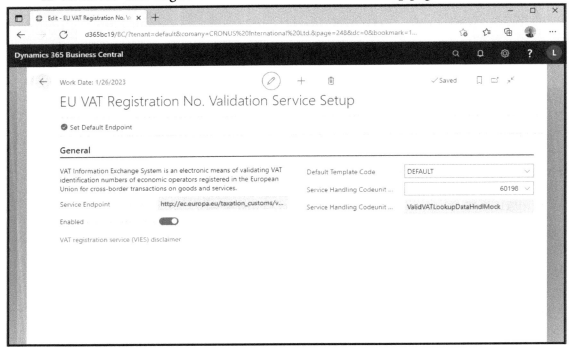

Figure 13.11 – EU VAT Registration No. Validation Service Setup with the additional fields Service Handling Code ID and Service Handling Code Name

You might want to have a look at the full context in codeunit VAT Reg. Log Mgt. Setup (60159): `https://github.com/PacktPublishing/Automated-Testing-in-Microsoft-Dynamics-365-Business-Central-Second-Edition/blob/main/Chapter 13 (VAT Registration No. Validation)/src/codeunit/VATRegLogMgtSetup.Codeunit.al`.

- With Subscriber:

In this case, we enable **DI** by providing an event publisher just before the Codeunit. Run call, enabling us to subscribe to it and fully handle the service interaction ourselves or, alternatively, providing the ID for a service handling codeunit to our choice:

```
OnBeforeCheckVIESForVATNoCodeunit(
```

```
      VATRegNoSrvCodeunitId, IsHandled);
  if IsHandled then
      exit;

  if VATRegNoSrvCodeunitId <> 0 then
    Codeunit.Run(VATRegNoSrvCodeunitId, VATRegistrationLog)
  else
    Codeunit.Run(
      Codeunit::"VAT Lookup Ext. Data Hndl",
        VATRegistrationLog);
```

Go and inspect the full context in codeunit VAT Reg. Log Mgt. Subscriber
(60169): https://github.com/PacktPublishing/Automated-Testing-
in-Microsoft-Dynamics-365-Business-Central-Second-Edition/
blob/main/Chapter 13 (VAT Registration No. Validation)/src/
codeunit/VATRegLogMgtSubscriber.Codeunit.al.

In a testing context, both cases facilitate us to provide a test double.

Notes

(1) In both From Setup and With Subscriber service handling options,
we make use of the interface behavior *avant la lettre* of Codeunit.Run by
dynamically providing a codeunit ID. As long as the called codeunit accepts the
VATRegistrationLog parameter, any codeunit ID will do. With the introduction
of the interface object type, we are able to implement **DI** in a more advanced
way. This is, however, now out of scope of this book. Also something to address in
the future. You might take advantage of the standard test examples referred to in the
Examples of standard tests with test doubles replacing external components section.

(2) Note the implementation of the **guard pattern**: in case no codeunit ID has been
given – that is when it equals 0 – the ID of the codeunit used in the standard code is
provided. See https://en.wikipedia.org/wiki/Guard_(computer_
science).

As said, to enable all three scenarios in the same AL extension, the Service Handling
Type field – and its Enum – was introduced. This leads to the following implementation of the
VATRegistrationValidation method:

```
local procedure VATRegistrationValidation()
var
  VATRegNoSrvConfig: Record "VAT Reg. No. Srv Config";
begin
  case "Service Handling Type" of
```

```
"Service Handling Type"::"Default Codeunit":
  VATRegValidationDefault();
"Service Handling Type"::"From Setup":
  begin
    VATRegNoSrvConfig.Get();
    VATRegNoSrvConfig.TestField("Handling Codeunit ID");
    VATRegValidationFromSetup(
      VATRegNoSrvConfig."Handling Codeunit ID");
  end;
"Service Handling Type"::"With Subscriber":
  begin
    VATRegValidationWithSubscriber();
  end;
  end;
end;
```

`VATRegValidationDefault` calls the standard code to interact with the hardcoded service as discussed above, `VATRegValidationFromSetup` injects the service codeunit ID taken from the setup, and `VATRegValidationWithSubscriber` calls the code that enables us to inject the dependency through an event publisher.

Using From Setup or With Subscriber?

Now that I gave you two options for **DI** to get the code testable, the question is which of the two is preferable. Or maybe broader: what are the pros and cons of either of them?

The *least intrusive* solution is performing DI by using `With Subscriber`. This one prevents having to update your code in potentially a lot of places contrary to `From Setup`. With the latter, the codeunit ID needs to be handed over from the place where the feature is triggered all the way down to where **DI** should happen. This can become very inefficient as this might lead to a long chain of handing over this value top-down.

As such, the `With Subscriber` option is eventually also the *simplest to implement* solution. It has however one drawback and that is the fact that with the subscribers, it is not always easy to reconstruct and understand the code flow.

> **Note**
> You might have noticed that our subscriber also makes use of the codeunit ID set on the setup. This, however, is just an easy solution for performing our **DI**: setting the ID for our test double.

Test code

Now, how do we write test code for this? Like always, let's first get our ATDD scenarios defined, which leads to the following four scenarios:

```
[SCENARIO #0001]  Test valid VAT registration number
                  with DI from Setup
[SCENARIO #0003]  Test valid VAT registration number
                  with DI using Subscriber
[SCENARIO #0002]  Test invalid VAT registration number
                  with DI from Setup
[SCENARIO #0004]  Test invalid VAT registration number
                  with DI using Subscriber
```

You can find the full ATDD scenarios with GIVEN-WHEN-THEN details in the Excel sheet `VAT Registration No. Validation - Chapter 13.xlsx` on GitHub: `https://github.com/PacktPublishing/Automated-Testing-in-Microsoft-Dynamics-365-Business-Central-Second-Edition/blob/main/Excel sheets/ATDD Scenarios/LookupValue - Chapter 13.xlsx`.

At the highest level, we do get the following four test functions:

```
[Test]
[HandlerFunctions(
  'VATRegistrationLogDetailsPageHandler')]
procedure TestValidVATRegistration
          NumberWithDIFromSetup()
// [FEATURE] VAT Registration No. Validation
begin
  // [SCENARIO #0001] Test valid VAT registration
  //                  number with DI from Setup
  TestValidVATRegistrationNumberWithDI(
    "Service Handling Type"::"From Setup")
end;

[Test]
[HandlerFunctions(
  'VATRegistrationLogDetailsPageHandler')]
procedure TestValidVATRegistrationNumberWithDIUsingSubscriber()
```

```
// [FEATURE] VAT Registration No. Validation
var
    VATRegLogMgtEvents: Codeunit "VAT Reg. Log Mgt. Events";
begin
    // [SCENARIO #0003] Test valid VAT registration
    //                  number with DI using Subscriber
    BindSubscription(VATRegLogMgtEvents);
    TestValidVATRegistrationNumberWithDI(
        "Service Handling Type"::"With Subscriber");
    UnbindSubscription(VATRegLogMgtEvents);
end;

[Test]
[HandlerFunctions('NoMatchFoundMessageHandler')]
procedure TestInvalidVATRegistrationNumberWithDIFromSetup()
// [FEATURE] VAT Registration No. Validation
begin
    // [SCENARIO #0002] Test invalid VAT registration
    //                  number with DI from Setup
    TestInvalidVATRegistrationNumberWithDI(
        "Service Handling Type"::"From Setup")
end;

[Test]
[HandlerFunctions('NoMatchFoundMessageHandler')]
procedure TestInvalidVATRegistrationNumberWithDIUsingSubscriber()
// [FEATURE] VAT Registration No. Validation
var
    VATRegLogMgtEvents: Codeunit "VAT Reg. Log Mgt. Events";
begin
    // [SCENARIO #0004] Test invalid VAT registration
    //                  number with DI using Subscriber
    BindSubscription(VATRegLogMgtEvents);
    TestInvalidVATRegistrationNumberWithDI(
        "Service Handling Type"::"With Subscriber");
    UnbindSubscription(VATRegLogMgtEvents);
end;
```

Notice how we bind – and subsequently unbind – the subscriber containing codeunit
VAT Reg. Log Mgt. Events (60160) specifically set up for our coded tests. Having
set the EventSubscriberInstance to Manual on this codeunit gives us precise
control over when the event it contains can be run. Note how the event subscriber
OnBeforeCheckVIESForVATNoCodeunit sets the VATRegNoSrvCodeunitId – if not
set yet – to the codeunit ID as has been configured in the **EU VAT Registration No. Validation
Service Setup**, with source table VAT Reg. No. Srv Config.

```
codeunit 60160 "VAT Reg. Log Mgt. Events"
{
   EventSubscriberInstance = Manual;

   [EventSubscriber(ObjectType::Codeunit,
       Codeunit::"VAT Reg. Log Mgt. Subscriber",
       'OnBeforeCheckVIESForVATNoCodeunit', '',
       false, false)]
   local procedure OnBeforeCheckVIESForVATNoCodeunit(
     var VATRegNoSrvCodeunitId: Integer;
     var IsHandled: Boolean)
   var
     VATRegNoSrvConfig: Record "VAT Reg. No. Srv Config";
   begin
     if IsHandled then
       exit;

     if VATRegNoSrvCodeunitId = 0 then begin
       VATRegNoSrvConfig.Get();
       VATRegNoSrvConfig.TestField("Handling Codeunit ID");
       VATRegNoSrvCodeunitId :=
         VATRegNoSrvConfig."Handling Codeunit ID";
     end;
   end;
}
```

The implementation of the four test functions has been abstracted in the two following procedures:

```
local procedure TestValidVATRegistrationNumberWithDI(
   ServiceHandlingType: enum "Service Handling Type")
var
   VATRegistrationNoEntryNo: Code[20];
```

```
begin
  // [GIVEN] EU country
  // [GIVEN] Well formatted VAT registration number
  // [GIVEN] Enabled service
  Initialize();
  // [GIVEN] Mock service handling codeunit returning
  //         valid
  SetMockServiceHandlingCodeunit(
    Codeunit::ValidVATLookupDataHndlMock);

  // [WHEN] Set and validate country and VAT
  //        registration number with DI
  VATRegistrationNoEntryNo :=
    SetAndValidateCountryAndVATRegistrationNumber(
      CountryCode, VATRegNoTxt, ServiceHandlingType);

  // [THEN] Validation details page is shown
  // [THEN] Validation details contains name and
  //        address
  // Handled by VATRegistrationLogDetailsPageHandler
  // [THEN] Valid VAT registration log entry
  VerifyValidVATRegistrationLogEntry(VATRegistrationNoEntryNo);
end;

local procedure
  TestInvalidVATRegistrationNumberWithDI(
    ServiceHandlingType: enum "Service Handling Type")
var
  VATRegistrationLog: Record "VAT Registration Log";
  VATRegistrationNoEntryNo: Code[20];
begin
  // [GIVEN] EU country
  // [GIVEN] Well formatted VAT registration number
  // [GIVEN] Enabled service
  Initialize();
  // [GIVEN] Mock service handling codeunit returning
  //         invalid
```

```
SetMockServiceHandlingCodeunit(
  Codeunit::InvalidVATLookupDataHndlMock);

// [WHEN] Set and validate country and VAT
//        registration number
VATRegistrationNoEntryNo :=
  SetAndValidateCountryAndVATRegistrationNumber(
    CountryCode, VATRegNoTxt, ServiceHandlingType);

// [THEN] No match found message thrown
// Handled by NoMatchFoundMessageHandler
// [THEN] Invalid VAT registration log entry
VerifyInValidVATRegistrationLogEntry(VATRegistrationNoEntryNo);
end;
```

Assuming it will not be too big a task to understand the details of these two functions, I would want to pick out one major detail in both functions: the last **GIVEN** where the "mock service handling codeunit" is set. Here is where the **DI** is prepared, setting the service handling codeunit ID on the setup. In the first function, this is set to `Codeunit::ValidVATLookupDataHndlMock`, and in the second one, it is set to `Codeunit::InvalidVATLookupDataHndlMock`. Both codeunits are the test doubles for these four automated tests and actually are not mocks but **stubs** as they do not contain any logic but just prepare a *valid* – codeunit `ValidVATLookupDataHndlMock` (60198) – or *invalid* – codeunit `InvalidVATLookupDataHndlMock` (60199) – response, which can only be used for that test it is designed for. Go to their objects files on GitHub here:

- `ValidVATLookupDataHndlMock`:

  `https://github.com/PacktPublishing/Automated-Testing-in-Microsoft-Dynamics-365-Business-Central-Second-Edition/blob/main/Chapter 13 (VAT Registration No. Validation)/test/ValidVATLookupDataHndlMock.Codeunit.al`

- `InvalidVATLookupDataHndlMock`:

  `https://github.com/PacktPublishing/Automated-Testing-in-Microsoft-Dynamics-365-Business-Central-Second-Edition/blob/main/Chapter 13 (VAT Registration No. Validation)/test/InvalidVATLookupDataHndlMock.Codeunit.al`

> **Note**
>
> A **stub** is a test double with no logic. It returns a value or gives a response specific to the tested context. Therefore, it is not reusable in another test context where another return value or response is needed. So, contrary to a **mock**, it does not hold a stand-in implementation of the service's logic.

Test execution

How about executing these four tests? Looking good: **green**!

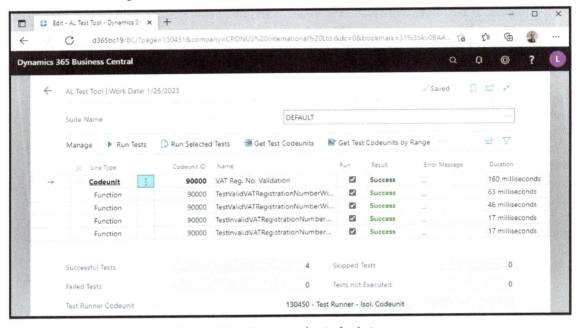

Figure 13.12 – Test example 13 after being run

> **Note**
>
> As the VAT registration number validation service handling is heavily making use of XMLDOM, these tests can only be run in an on-prem environment.

Summary

In this chapter, you learned first how to test incoming calls, that is, how to test your own APIs, and secondly, how to test outgoing calls, that is, how to enable the usage of **test doubles** by means of **dependency injection** (**DI**) enabling in your code.

With this chapter, the book has come to an end. But your journey into test automation might have just started. To me, every time it's still fun to pick up. It enables me to go astray in my application code, finding the results of my test pulling me back. It allows me to refactor my code at any time. It also enables me to reflect on the quality of the code more easily because it can be changed and instantly checked. I could go on with my praising and more or less redo the first chapter of this book but I won't, as it is time to conclude for now. I hope you enjoyed reading and practicing with me as much as I did while writing.

Section 6: Appendix

The last section of this book pays attention to setting up your environment to get you up and running with Business Central, VS Code, and AL development, and the code examples to be found in the repository on GitHub.

This section contains the following chapter:

- *Appendix, Getting Up and Running with Business Central, VS Code, and the GitHub Project*

Appendix
Getting Up and Running with Business Central, VS Code, and the GitHub Project

This is a book about writing automated tests in Microsoft Dynamics 365 Business Central, not about how to develop extensions with VS Code and the AL language. It is assumed that you know how to use VS Code in conjunction with the AL development language and how Dynamics 365 Business Central works as a platform and an application. Based on this, we dive straight into the code in *Chapter 2*, *Test Automation and Test-Driven Development*, without explaining anything about the development tools we are using, and continue to do that in the following chapters.

In this appendix, however, we pay some attention to setting up your Business Central environment and VS Code, and the code examples to be found in the repository on GitHub.

This appendix covers the following topics:

- Setting up your Business Central environment
- The GitHub repository
- Setting up VS Code
- Notes on the AL code

Setting up your Business Central environment

While the writing of app and test code is done in VS Code and could be done to a certain extent without a Business Central environment, you always eventually need to deploy it to Business Central. The proof of the pudding is in the eating, right? You want to see that it can be published and installed and, even more, you want to verify that your app code passes the checks your test code has been set up to do.

For development purposes, it's often best to have a local installation of Business Central or one in a VM for your own use. For the test automation worked on in this book, it has to be an on-prem installation – that is, not using the SaaS version – as some of the standard Microsoft test apps and test libraries can only run on-prem. Whether you're using a VM for this or not is of no importance here.

An on-prem installation can either be achieved by installing it from a product DVD or by using **BcContainerHelper**, which will be discussed next.

Setting up Business Central from a product DVD

If you are going to set up Business Central from a product DVD, take the following steps:

1. Select the**Advanced installation options**.

Figure A.1 – Select Advanced installation options

2. On the next screen, select the **Choose an installation option**.

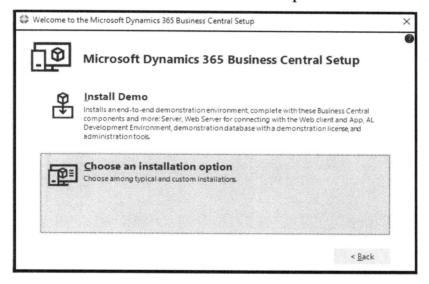

Figure A.2 – Select Choose an installation option

3.   On the next screen, select **Developer | Customize....**

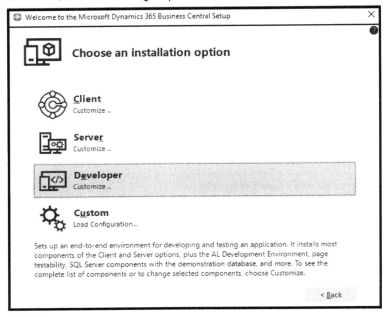

Figure A.3 – Select Developer > Customize

4.   On the next screen, verify that the following options have been selected and select **Next**.

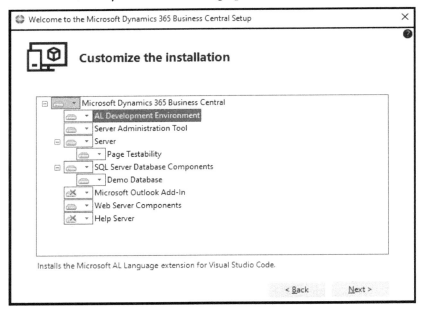

Figure A.4 – Select Next

On any next screen that follows, you just select **Next**.

> **Note**
>
> To download a Business Central product DVD, use the following URL, replacing `<version>` with the relevant internal version, and follow the instructions under the *Download update package* section: `https://docs.microsoft.com/en-us/dynamics365/business-central/dev-itpro/upgrade/upgrading-cumulative-update-v<version>`. You typically would like to download – and install – the worldwide version, aka w1.

Getting standard test and test library apps deployed

A product DVD-based installation leaves you unfortunately with some additional work to do as it does not include the deployment of any of the standard test and test library apps. You need to get that done by publishing (and installing) the apps from the following directories on the product DVD:

- `\Applications\TestFramework\TestRunner`, containing the AL Test Tool

- `\Applications\TestFramework\TestLibraries`, containing the new standard test library apps

- `\Applications\System Application\Test`, containing the standard system application test and test library apps

- `\Applications\BaseApp\Test`, containing the standard application test and test library apps

As these are altogether quite a number of apps to deploy, I have created a PowerShell script that will do that for you in one go on your on-prem Business Central installation, given you have downloaded the product DVD. You can find the script here: `https://github.com/PacktPublishing/Automated-Testing-in-Microsoft-Dynamics-365-Business-Central-Second-Edition/blob/main/Scripts/DeployTestAppsFromDVD.ps1`.

Setting up Business Central in a container

Instead of installing Business Central directly on your system, you could also do this in a docker container using, for example, the **BcContainerHelper** PowerShell module. With a containerized Business Central installation you take the advantage of having it easily set up and refreshed again when needed.

In order to be able to do this, the following prerequisites should be installed on your system first:

- A docker engine, for example, docker desktop. Find info here on installing docker on Windows 10: `https://docs.docker.com/docker-for-windows/install/`.
- The PowerShell **BcContainerHelper** module itself. Instructions can be found at `https://www.powershellgallery.com/packages/BcContainerHelper`.

Once that's done, use the following script to create a new Business Central container. In this case, a w1 2021 wave 2 container (internal version 19). Note that the highlighted parts need to be replaced by your specifics:

```
Install-Module -Name bccontainerhelper -force

# set accept_eula to $true to accept the eula found here:
#     https://go.microsoft.comfwlink/?linkid=861843
$accept_eula = $true

# set container name and version
$containerName = 'your-container-name'

# set artifact url
$artifactUrl = Get-BCArtifactUrl -country "base" `
                                 -version "19"

# set credentials for UserPassword usage
$login = 'your-user-name'
$password = 'your-password'
$securepassword = (ConvertTo-SecureString `
                -String $password OnPrem `
                -AsPlainText OnPrem `
                -Force)
$credential = New-Object PSCredential(`
```

```
                $login, $securepassword)

# create container
New-BCContainer -accept_eula:$accept_eula `
                -containerName $containerName `
                -artifactUrl $artifactUrl `
                -auth UserPassword `
                -credential $credential `
                -memoryLimit:10GB `
                -updateHost `
                -shortcuts StartMenu `
                -isolation process `
                -includeTestToolkit
```

The script can be found here: `https://github.com/PacktPublishing/Automated-Testing-in-Microsoft-Dynamics-365-Business-Central-Second-Edition/blob/main/Scripts/NewBCContainer.ps1`.

The includeTestToolkit parameter

The `includeTestToolkit` parameter in the `New-BCContainer` call will make sure all standard apps that relate to test automation will be installed: test apps, test library apps, and the Test Runner. If you do not want all standard test apps, add the `includeTestLibrariesOnly` parameter.

> **Notes**
>
> (1) The script sets up Business Central to use user passwords. This is because this will always work. When using Windows authentication, this might not always be the case if, for example, your laptop is not part of a domain and you are using your Microsoft account to log in.
>
> (2) Note that the country setting "base" creates a w1 Cronus demo data company.

The GitHub repository

The various code examples in this book have been uploaded to a dedicated GitHub repository. This repository can be accessed through this link: `https://github.com/PacktPublishing/Automated-Testing-in-Microsoft-Dynamics-365-Business-Central-Second-Edition`.

Structure of the GitHub repository

By following the previous link, you will land on the main page of this repository, showing you the following folders:

- `Chapter 02`
- `Chapter 03`
- `Chapter 06 (LookupValue Extension)`
- `Chapter 07 (LookupValue Extension)`
- `Chapter 08 (LookupValue Extension)`
- `Chapter 10 (LookupValue Extension)`
- `Chapter 11 (LookupValue Extension)`
- `Chapter 12 (LookupValue Extension)`
- `Chapter 13 (LookupValue Extension)`
- `Chapter 13 (VAT Registration No. Validation)`
- `Excel Sheets`
- `Graphics`
- `LookupValue Extension`
- `Scripts`

In this section, we will discuss what the various folders contain. Notice that GitHub puts them in alphabetical order. This does not fully relate to the order in which the folders are used in the book. However, I will elaborate on them in the order of the book.

Main and BC18 branches

But before we step into the various folders, it needs to be mentioned that this repo consists of two branches: `main` and `BC18`. As the name of the latter branch indicates, the code that resides in this branch is to be used for Business Central 2021 wave 1, aka BC18, and might also be applied to Business Central 2020 wave 2, aka BC17. The code in the `main` branch is based on Business Central 2021 wave 2, aka BC19. The reason that these two branches are provided is because the customer template functionality has changed in BC19.

> **Notes**
>
> (1) Like almost any other GitHub repository, ours also contains a `.gitignore`, `LICENSE`, and `README.md` file.
>
> (2) Each VS Code folder discussed below includes the `app.json` file that allows you to deploy the code as an extension.

Chapter 02

This folder contains the code used to demo the usage of Test-Driven Development in Business Central in *Chapter 2, Test Automation and Test-Driven Development.*

Chapter 03

This folder contains the code examples from *Chapter 3, The Testability Framework,* including the `MyTestsExecutor` page that allows you to run the code examples.

Excel Sheets

This folder holds the following two subfolders:

- `ATTD Scenarios`
- `Examples`

In the first folder, `ATTD Scenarios`, you will find two Excel files:

- `LookupValue.xlsx`: This file contains a listing of all **Acceptance Test-Driven Development** (ATDD) scenarios, that is, `GIVEN-WHEN-THEN`, detailing the full customer wish, as introduced in *Chapter 6, From Customer Wish to Test Automation – the Basics,* for our **LookupValue** extension. This is elaborated on further in *Chapter 6,* and the following.

- `LookupValue - Chapter 12.xlsx`: This file is a continuation of `LookupValue.xlsx` with a number of new scenarios added for *Chapter 12, Writing Testable Code.*

The `Examples` folder holds all the example Excel files that are discussed in *Chapter 5*, *Test Plan and Test Design*, and *Chapter 6*, *From Customer Wish to Test Automation – the Basics*. Next to that, it also holds the `Clean sheet.xlsx` file, which is a clean version that you can use to write your own ATDD scenarios.

Chapter 06 (LookupValue Extension)

This folder contains the completed test code examples from *Chapter 6* for the app code in the `src` folder. If you would like to work through all test examples yourself, make a copy of this folder and remove all objects from the `test` folder. This way you have a clean VS Code project folder only containing the app code and ready to start programming the test code.

> **Note**
> To make best use of the chapter folders, treat them as a workspace folder in VS Code, that is, open the chapter folder in VS Code. Do not open the whole repo as a workspace in VS Code.

Chapter 07 (LookupValue Extension)

Starting from the code in the `Chapter 06 (LookupValue Extension)` folder, this folder adds the completed code examples from *Chapter 7*, *From Customer Wish to Test Automation – Next Level*.

Chapter 08 (LookupValue Extension)

Starting from the code in the `Chapter 07 (LookupValue Extension)` folder, this folder adds the completed code examples from *Chapter 8*, *From Customer Wish to Test Automation – the TDD way*.

Chapter 09 (LookupValue Extension)

Starting from the code in the `Chapter 08 (LookupValue Extension)` folder, this folder adds the completed code examples from *Chapter 9*, *How to Integrate Test Automation in Daily Development Practice*.

Chapter 10 (LookupValue Extension)

Starting from the code in the `Chapter 09 (LookupValue Extension)` folder, this folder adds the completed code examples from *Chapter 10*, *Getting Business Central Standard Tests Working on Your Code*.

Note, however, that *Chapter 10* focuses on the Business Central standard tests so the test developed in the previous chapters has been thrown out, only retaining our test libraries.

Chapter 11 (LookupValue Extension)

Starting from the code in the `Chapter 09 (LookupValue Extension)` folder, this folder adds the completed code examples from *Chapter 11*, *How to Construct Complex Scenarios*. It does not include code from *Chapter 10*, *Getting Business Central Standard Tests Working on Your Code* as, this was a sidestep regarding the Microsoft standard tests.

Chapter 12 (LookupValue Extension)

Starting from the code in the `Chapter 11 (LookupValue Extension)` folder, this folder adds the completed code examples from *Chapter 12*, *Writing Testable Code*.

Chapter 13 (LookupValue Extension)

Starting from the application code in the `Chapter 12 (LookupValue Extension)` folder, this folder adds the completed code for *Test example 12 – testing incoming calls: Lookup Value API*, of *Chapter 13*, *Testing Incoming and Outgoing Calls*.

Scripts

This folder holds the PowerShell scripts discussed in this appendix plus one SQL script, `RemoveAllUserRelatedData.sql`, used in *Test example 12 – testing incoming calls: Lookup Value API*, of *Chapter 13*.

Chapter 13 (VAT Registration No. Validation)

This folder contains the code for *Test example 13 – VAT Registration No. validation*, of *Chapter 13*, *Testing Incoming and Outgoing Calls*.

LookupValue Extension

When releasing an extension, it is best practice to separate the application from the test code. Therefore, this folder contains separate application and test extensions based on the work done in *Chapter 6* and the following chapters, and, in addition, also the app and test code for the features that were not discussed in the book. We will elaborate somewhat more on this below in *How to structure your project with respect to app and test app?*.

Graphics

Find all original graphics used in the book in this folder.

Setting up VS Code

If you are new to extension development with VS Code and the AL language, you might want to practice this first. There are various resources out there, but a very comprehensive one, though not too voluminous, is *Dynamics 365 Business Central Development Quick Start Guide*, by Stefano Demiliani and Duilio Tacconi. Here is a link to the Packt page where you also can order the book: `https://www.packtpub.com/business/dynamics-365-business-central-development-quick-start-guide`.

> **Note**
> There are, of course, various other resources out there that can help you to get up to speed with AL development in VS Code. One of them is this comprehensive video, *Best practices in Visual Studio Code development*, `https://www.youtube.com/watch?v=u24hyetS30Y`, which also is to be found in this playlist: `https://www.youtube.com/playlist?list=PL1FESh9FqyhQxkZXYYp1X1tTpXf3-9iSR`.

VS Code project

In *Dynamics 365 Business Central Development Quick Start Guide*, Stefano Demiliani and Duilio Tacconi explain in a practical way how to set up a new extension project in VS Code, start programming in AL, and deploy the extension to Business Central. The projects you will find in the GitHub repository for this book have been created in the same way: as explained in the *The GitHub repository* section for each chapter there is a separate folder containing the AL code objects and the `app.json` file. If you have been working on extensions already, you will notice that one major resource is missing: the `launch.json` file.

launch.json

To be able to use a chapter folder in the GitHub repo as a complete VS Code AL extension project folder, you need to add a `launch.json` file to it. The `launch.json` file is typically stored in the `.vscode` subfolder of the project folder and contains information on the Business Central installation that the extension will be launched on. To get a `.vscode` folder with a `launch.json` file, follow these steps:

1. Open the relevant chapter folder in VS Code with the AL Language extension installed.

2. Use the AL: Download symbols command from the VS Code command palette. If no launch.json file is present in this chapter folder, VS Code will ask you in the command palette to choose a server, either **Microsoft cloud sandbox** or **Your own server**.

3. Select **Your own server** as your Business Central installation will be an on-prem one – see *Setting up your Business Central environment* section. VS Code will now create a launch.json file in the .vscode folder.

4. Your launch.json file might look somewhat like this (when updated so it complies with your Business Central environment):

```
{
  "version": "0.2.0",
  "configurations": [
    "name": "bc19",
    "request": "launch",
    "type": "al",
    "environmentType": "OnPrem",
    "server": "http://d365bc19",",
    "serverInstance": "BC",
    "port": 7049,
    "tenant": "default",
    "authentication": "UserPassword",
    "schemaUpdateMode": "Synchronize",
    "startupObjectType": "Page",
    "startupObjectId": 130451,
    "breakOnError": true,
    "launchBrowser": false,
    "enableLongRunningSqlStatements": true,
    "enableSqlInformationDebugger": true,
    "tenant": "default"
  ]
}
```

> **Note**
>
> Compared to the default launch.json file, you will note that I have added/updated a couple of useful keywords, such as startupObjectId, the start up page 130451, which is the AL Test Tool page.
>
> You will find more details in *Dynamics 365 Business Central Development Quick Start Guide,* by Stefano Demiliani and Duilio Tacconi.

app.json

The app.json file, also called the manifest, defines the meta description of the extension and has been provided in each chapter folder already on GitHub. Here is the one for *Chapter 11, How to Construct Complex Scenarios*, only listing the relevant parts:

```json
{
  "id": "e26890f8-fafe-49c6-8951-2c1457921f9b",
  "name": "LookupValue",
  "publisher": "fluxxus.nl",
  "brief": "LookupValue extension for book Automated
      Testing in Microsoft Dynamics 365 Business
      Central",
  "description": "LookupValue extension as basis to
      test examples in chapter 6 and following for the
      book Automated Testing in Microsoft Dynamics 365
      Business Central written by Luc van Vugt and
      published by Packt",
  "version": "2.0.0.0",
  "dependencies": [
    {
      "id": "e7320ebb-08b3-4406-b1ec-b4927d3e280b",
      "publisher": "Microsoft",
      "version": "18.0.0.0",
      "name": "Any"
    },
    {
      "id": "dd0be2ea-f733-4d65-bb34-a28f4624fb14",
      "publisher": "Microsoft",
      "version": "18.0.0.0",
      "name": "Library Assert"
    },
    {
      "id": "5095f467-0a01-4b99-99d1-9ff1237d286f",
      "publisher": "Microsoft",
      "version": "18.0.0.0",
      "name": "Library Variable Storage"
    },
    {
      "id": "5d86850b-0d76-4eca-bd7b-951ad998e997",
      "publisher": "Microsoft",
```

```
         "version": "18.0.0.0",
         "name": "Tests-TestLibraries"
      },
      {
         "id":   "23de40a6-dfe8-4f80-80db-d70f83ce8caf",
         "name":   "Test Runner",
         "publisher":   "Microsoft",
         "version":   "18.0.0.0"
      }
   ],
   "platform": "18.0.0.0",
   "application": "18.0.0.0",
   "idRanges": [
      {
         "from": 50000,
         "to": 50099
      },
      {
         "from": 80000,
         "to": 81099
      }
   ],
   "target": "OnPrem",
   "showMyCode": true,
   "runtime": "7.0"
}
```

app.json explained

I reckon a couple of notes on some specific keywords of the app.json file will help you to have a better understanding:

- version has been set to 2.0.0.0 as this is the second version of the **LookupValue** extension. The first version was created with the first edition of this book. See https:// github.com/PacktPublishing/Automated-Testing-in-Microsoft-Dynamics-365-Business-Central.

- In dependencies, you can find the Microsoft extensions needed to create the tests where we make use of various standard helper functions. More details can be found in the chapter where one or more of these extensions are needed.

- Setting platform and application to 18.0.0.0 ensures that the app.json can be applied to BC18 and beyond, so also, BC19.

- For all chapters, `idRanges` contains two ranges: 50000 to 50099 for the *app code*, and 80000 to 81099 for the *test code*.

- `target` is set to `OnPrem` as some standard tests – see *Chapter 10* – have been set to `Scope OnPrem` and permission testing – see *Chapter 8* – can only be done on an on-prem Business Central installation.

Creating your own app.json

Instead of using the `app.json` file, as provided on GitHub, you might create your own `app.json` file, like you get when creating a newly created VS Code AL project. If you do that, make sure that your extension has a different (and unique) name and ID, as I already, while working on this book, deployed the extension **LookupValue** a zillion times.

> **Note**
>
> In **Tip #10** of the *Top tips and tricks on how to create Per-tenant extensions* blog (`https://simplanova.com/top-tips-tricks-per-tenant-extensions/`), Dmitry Katson explains what happens if your extension is not unique and how to make it unique.

Setting dependencies on the Microsoft test and test library apps

In the `app.json` file above, you can see how dependencies on the Microsoft test library apps have been defined. It can be quite a nuisance to find the exact details of the Microsoft test library apps you need to define a dependency on for your (test) app. It's not a difficult exercise, but the info is not always readily available. Because of that, I wrote a **note to self** on my blog that lists all Microsoft standard test and test library apps with their details: `https://www.fluxxus.nl/index.php/bc/test-apps-dependencies/`.

settings.json

A `.vscode` folder is provided containing a `settings.json` file defining a number of settings for **waldo's CRS AL Language** extension that determine the standard file-naming pattern, helping you – and me! – with an easy renaming of the object files using the `CRS: Rename` command.

> **Note**
>
> To learn more about **waldo's CRS AL Language** extension here: `https://marketplace.visualstudio.com/items?itemName=waldo.crs-al-language-extension`.

How to structure your project with respect to app and test app?

As already discussed above, it is best practice – when releasing an extension – to separate the application from the test code. For `AppSource`, this is an implicit requirement as any dependencies on the test framework or test libraries will not get through validation.

While working on app and test code for a new extension, I prefer to have both in one project; this allows an easy deployment irrespective of an update done to the app or test code. In both cases, I only have to deploy the same extension. This is exactly what I will do through the book.

Once the app code and its accompanying test code are ready, I split the single project into two and the test app will need to take a dependency on the application app. To make my work on both as efficient as possible from this moment on, I will set up a multi-root workspace file containing the app and test apps project folders.

The completed **LookupValue** extension code can be found in the `LookupValue Extension` folder on the GitHub repo discussed next. This does contain a multi-root workspace file.

> **Note**
>
> Read about multi-root workspace here: `https://code.visualstudio.com/docs/editor/workspaces#_multiroot-workspaces`

Notes on the AL code

Now, some final notes on the code on GitHub and the code examples in the book.

Prefix or suffix

When building your own extension, it is best practice to make use of a so-called "affix" (a 3–6-digit *prefix* or *suffix*) in the names of your objects, fields, and controls. Affixes are mandatory for AppSource extensions, and VAR or ISV partners must register an affix with Microsoft. We did choose not to use a prefix/suffix for the following reasons:

- The `LookupValue` extension is not a registered extension.

- The use of a prefix/suffix would not contribute to the readability of the code examples.

> **Note**
>
> More information on the usage of a *prefix* or *suffix* can be found here: `http://docs.microsoft.com/en-us/dynamics365/business-central/dev-itpro/compliance/apptest-prefix-suffix`.

Word wrap

When adding code examples to a book, there is always the challenge of how to neatly format long-lined statements to keep the code readable. In the code examples in this book, this has been done by enforcing *word wrap*, even though the code might not compile anymore. All of the code in the GitHub repository, however, is technically OK.

Index

`Packt.com`

Subscribe to our online digital library for full access to over 7,000 books and videos, as well as industry leading tools to help you plan your personal development and advance your career. For more information, please visit our website.

Why subscribe?

- Spend less time learning and more time coding with practical eBooks and Videos from over 4,000 industry professionals

- Improve your learning with Skill Plans built especially for you

- Get a free eBook or video every month

- Fully searchable for easy access to vital information

- Copy and paste, print, and bookmark content

Did you know that Packt offers eBook versions of every book published, with PDF and ePub files available? You can upgrade to the eBook version at `packt.com` and as a print book customer, you are entitled to a discount on the eBook copy. Get in touch with us at `customercare@packtpub.com` for more details.

At `www.packt.com`, you can also read a collection of free technical articles, sign up for a range of free newsletters, and receive exclusive discounts and offers on Packt books and eBooks.

Other Books You May Enjoy

If you enjoyed this book, you may be interested in these other books by Packt:

Microsoft Dynamics 365 Project Operations

Robert Houdeshell

ISBN: 9781801072076

- Configure key elements of Project Operations to drive improved collaboration with your customers
- Discover how Project Operations is interconnected with Microsoft 365 and Dynamics 365 Platform
- Understand the Project Opportunity-to-Quote-to-Contract workflow and its implications for selling
- Find out how to set up and utilize direct staffing and centralized staffing models
- Explore Project Timeline Management using Task, Board, and Timeline views
- Find out how information flows to finance and operations in Project Operations

Mastering Microsoft Dynamics 365 Business Central

Stefano Demiliani, Duilio Tacconi

ISBN: 9781789951257

- Create a sandbox environment with Dynamics 365 Business Central
- Handle source control management when developing solutions
- Explore extension testing, debugging, and deployment
- Create real-world business processes using Business Central and different Azure services
- Integrate Business Central with external applications
- Apply DevOps and CI/CD to development projects
- Move existing solutions to the new extension-based architecture

Programming Microsoft Dynamics 365 Business Central

Marije Brummel, David Studebaker, and Chris Studebaker

ISBN: 9781789137798

- Programming using the AL language in the Visual Studio Code development environment

- Explore functional design and development using AL

- How to build interactive pages and learn how to extract data for users

- How to use best practices to design and develop modifications for new functionality integrated with the standard Business Central software

- Become familiar with deploying the broad range of components available in a Business Central system

- Create robust, viable systems to address specific business requirements

Packt is searching for authors like you

If you're interested in becoming an author for Packt, please visit `authors.packtpub.com` and apply today. We have worked with thousands of developers and tech professionals, just like you, to help them share their insight with the global tech community. You can make a general application, apply for a specific hot topic that we are recruiting an author for, or submit your own idea.

Share Your Thoughts

Now you've finished *Automated Testing in Microsoft Dynamics 365 Business Central*, we'd love to hear your thoughts! Scan the QR code below to go straight to the Amazon review page for this book and share your feedback or leave a review on the site that you purchased it from.

https://packt.link/r/1801816425

Your review is important to us and the tech community and will help us make sure we're delivering excellent quality content.